SpringerBriefs in Molecular Science

Green Chemistry for Sustainability

Series editor

Sanjay K. Sharma, Jaipur, India

More information about this series at http://www.springer.com/series/10045

Huanfeng Jiang · Chuanle Zhu
Wanqing Wu

Haloalkyne Chemistry

 Springer

Huanfeng Jiang
School of Chemistry and Chemical
  Engineering
South China University of Technology
Guangzhou
China

Wanqing Wu
School of Chemistry and Chemical
  Engineering
South China University of Technology
Guangzhou
China

Chuanle Zhu
School of Chemistry and Chemical
  Engineering
South China University of Technology
Guangzhou
China

ISSN 2191-5407   ISSN 2191-5415   (electronic)
SpringerBriefs in Molecular Science
ISSN 2212-9898
SpringerBriefs in Green Chemistry for Sustainability
ISBN 978-3-662-48999-4   ISBN 978-3-662-49001-3   (eBook)
DOI 10.1007/978-3-662-49001-3

Library of Congress Control Number: 2015955865

Springer Heidelberg New York Dordrecht London

Printed on acid-free paper

Springer-Verlag GmbH Berlin Heidelberg is part of Springer Science+Business Media
(www.springer.com)

Preface

Green method, atom economy, and concise synthesis are terms frequently mentioned in chemistry-related publications and presentations nowadays. Inspired by the requirements of green and sustainable chemistry, considerable efforts have been devoted to developing general and practical methods to construct complex molecules by taking advantages of chemical reagents with diverse and tunable reactivity. Haloalkynes, such as bromoalkynes, chloroalkynes, and iodoalkynes, are a significant class of molecules that have these futures and been widely utilized in organic synthesis.

This book summarizes the general methods to prepare haloalkyne reagents and also presents the selected examples to highlight the progress on the development and applications of convenient and concise synthetic approaches involving haloalkynes. According to the reactive sites of haloalkynes involved in the transformations, we classify these reactions into three types: (i) the transformations of carbon–halo bond motif; (ii) the diverse functionalization of carbon–carbon triple bond unit; and (iii) the reactions involving both carbon–halo bond motif and carbon–carbon triple bond unit. The emphasis is put on the reaction mechanism aspects and the synthetic utilities of the obtained products.

The primary purpose of this book is to illustrate the diverse reactivities and applications of haloalkyne reagents, to describe the experimental techniques of these valuable transformations in detail, as well as to enlighten the researchers to answer the unsolved problems in haloalkyne chemistry. This book should be useful to researchers in organic and organometallic chemistry as well as catalysis from both academia and industry. Significantly, doctorate students and postdoctoral researchers should be motivated by these innovations in chemistry.

I am especially grateful to the cooperative contributions made by all the authors. Without their efforts and expertise, this book would not have been possible. I would also like to thank the organizational support from Springer to overcome the troubles encountered in the production of this book.

Guangzhou, China   Huanfeng Jiang
September 2015

Contents

Abbreviations

Ac	Acetyl
Ad	Adamantyl
BINAP	2,2′-Bis(diphenylphosphino)-1,1′-binaphtyl
Bn	Benzyl
BQ	Benzoquinone
Bu	Butyl
COD	Cyclooctadiene
Cp	Cyclopentadienyl
Cy	Cyclohexyl
DABCO	1,4-Diazabicyclo[2.2.2]octane
dba	Dibenzylideneacetone
DCE	1,2-Dichloroethane
DCM	Dichloromethane
DMAP	4-Dimethylaminopyridine
DMEDA	N,N'-Dimethyl-1,2-ethanediamine
DMF	Dimethylformamide
DMSO	Dimethyl sulfoxide
Equiv	Equivalent(s)
Et	Ethyl
h	Hour(s)
HMPT	Hexamethyl phosphoryl triamide
ILs	Ionic liquids
iPr	Isopropyl
NBS	N-Bromosuccinimide
NHC	N-Heterocyclic carbene
NIS	N-Iodosuccinimide
Nu	Nucleophile
OTf	Trifluoromethanesulfonic
PE	Petroleum ether
Ph	Phenyl
Py	Pyridyl

1,10-phen	1,10-Phenanthroline
rt	Room temperature
scCO$_2$	Supercritical carbon dioxide
Sia	(Secondary)isoamyl
TBAB	Tetrabutylammonium bromide
TBAF	Tetrabutylammonium fluoride
TBHP	*tert*-Butyl hydroperoxide
TBS	*tert*-Butyldimethylsilyl
TEA	Triethyl amine
TEMPO	2,2,6,6-Tetramethylpiperidine 1-oxyl
THF	Tetrahydrofuran
TIPS	Triisopropylsilyl
TLC	Thin-layer chromatography
TMEDA	Tetramethylethylenediamine
TMS	Trimethylsilyl
Tol	4-Methylphenyl

Chapter 1
Introduction

Abstract Inspired by the demand of green and sustainable chemistry, modern synthetic chemists have devoted to develop general and practical methods to construct complex molecules. Due to the *sp* hybridization of the triple bond and the connected halogen atom, haloalkynes, such as bromoalkynes, chloroalkynes and iodoalkynes, have shown both controllable electrophilic and nucleophilic properties, rendering them highly versatile and robust synthons. As the immense usefulness of haloalkynes, impressive efforts have been devoted to this area in the past decades and many novel chemical reactions have been developed. In this chapter, we will introduce the physical property of haloalkynes, classify the reaction intermediate types derived from haloalkynes, and also hope to give the readers a comprehensive understanding of haloalkyne chemistry.

Keywords Green and sustainable chemistry · General and practical methods · Haloalkyne chemistry · Controllable electrophilic and nucleophilic properties · Versatile and robust synthons

The development of efficient and practical synthetic methods upon readily available reagents to construct molecular complexity has greatly accelerated the advancement of synthetic chemistry and related subjects. Inspired by the demand of green and sustainable chemistry, modern synthetic chemists have devoted to develop general and practical methods to construct complex molecules, as well as maximizing atom economy and minimizing synthetic steps [1]. During the past few decades, considerable progress has been achieved to fulfil these goals by taking advantages of chemical reagents with diverse and tunable reactive properties. Among them, haloalkynes, such as bromoalkynes, chloroalkynes and iodoalkynes, are a significant class of molecules that have been widely utilized in organic synthesis [2].

Generally, haloalkynes, especially iodoalkynes, are good Lewis acids. In 1981, Laurence and co-workers have demonstrated that the Lewis acidity of haloalkynes could affect the vibrational spectra of these compounds [3]. In 2000, Goroff and co-workers reported an unusual solvent effect, in which the solvent could significantly change the ^{13}C NMR chemical shift of iodoalkynes **1** and **2** (Table 1.1) [4].

© The Author(s) 2016
H. Jiang et al., *Haloalkyne Chemistry*, SpringerBriefs in Green Chemistry for Sustainability, DOI 10.1007/978-3-662-49001-3_1

Table 1.1 ^{13}C NMR chemical shifts of **1** and **2** (in ppm)

A B C I━━━━━━━I **1**		A B C D I━━━━━━━━━━I **2**	
Compound **1** in	δ(A)	δ(B)	δ(C)
CDCl$_3$	0.9	78.5	59.7
DMSO-d_6	14.6	76.3	58.8
Compound **2** in	δ(A)	δ(B)	δ(C, D)
CDCl$_3$	1.9	78.8	58.8, 62.0
DMSO-d_6	17.9	77.4	58.3, 62.7

Later, they confirmed that it was a general phenomenon. The chemical shift of C-1 in 1-iodo-2-phenylethyne (**3**) is 6.2 ppm in CDCl$_3$, but moves to 17.7 ppm in DMSO-d_6 and 19.4 ppm in pyridine-d_5 (Table 1.2) [5]. Computational evidences indicated that this solvent effect came directly from polarization of iodoalkyne triple bond in a Lewis acid-base complex with the solvent. It could predict that an increase in the electron density at C-1 would lead to a decrease in chemical shift.

Due to the *sp* hybridization of the triple bond and the connected halogen atom, haloalkynes show both controllable electrophilic and nucleophilic properties, rendering them highly versatile and robust synthons. Traditionally, haloalkyne reagents are served as a source of acetylides through metal-halogen exchange (Scheme 1.1, A). Until 1943, Ott [6] disclosed that haloalkyne derivatives could also be employed as equivalents of electrophilic acetylenic moiety, which would go through an addition-elimination procedure upon the reaction with nucleophiles. Importantly, the first enantioselective version was realized by Jørgensen [7] with the treatment of a chiral phase-transfer catalyst in 2006 (Scheme 1.1, B). Additionally, Boger [8] and Gevorgyan [9] reported respectively that haloalkynes could serve as effective sources of the corresponding X$^+$ ion or both X$^+$ and acetylide ions in the presence of organolithium species (Scheme 1.1, C and D). Noteworthy, as the continuous efforts of Jiang's group, the potential reactive abilities of haloalkyne reagents became fully apparent along with the development

Table 1.2 ^{13}C NMR chemical shifts of **3** formed Lewis acid-base complexes (in ppm)

A Ph━━━I ← O=S(Me)(Me) **3**	A Ph━━━I ← N(pyridine) **3**
Compound **3** in	δ(A)
CDCl$_3$	6.2
DMSO-d_6	17.7
pyridine-d_5	19.4

of transition metal catalysis [2]. Generally, under the treatment of transition metal catalysts, haloalkynes can be deemed as a dual functionalized molecule. Depending on reaction conditions, several reaction intermediates, such as σ-acetylene-metal complex (Scheme 1.1, Type I), π-acetylene complex (Scheme 1.1, Type II) and halovinylidene-metal complex (Scheme 1.1, Type III) can be formed and undergo further transformations to construct various of useful compounds. Additionally, the reactive halogen substituents can be further functionalization, which permits the rapid assembly of structural complexity. As the immense usefulness of haloalkynes, impressive efforts have been devoted in this area in the past decades and many novel chemical reactions have been developed [10].

In this book, we will classify the general methods to prepare haloalkyne reagents and also present selected examples to highlight the progress on the development and applications of convenient and concise synthetic approaches involving haloalkyne reagents. The designed methods, as well as serendipitous observations will be discussed with special emphasis on the mechanistic aspects and the synthetic utilities of the obtained products, aiming to illustrate the potential applications of haloalkyne chemistry in a wide spectrum of fields, including natural-product synthesis, materials science, and bioorganic chemistry. Importantly, the general procedure for each transformation of haloalkynes is described in detail. This book should be useful to researchers in organic and organometallic chemistry as well as catalysis both from academia and industry. Significantly, doctorate students, postdoctoral researchers and young researchers should be motivated by these innovations in chemistry. We hope this book could not only draw the blueprint of haloalkyne chemistry, and help the readers to comprehensively know and understand the diverse reactivities and applications of haloalkyne reagents, but also could be used as a handbook for researchers to develop novel catalytic systems to answer the unsolved challenges in haloalkyne chemistry and exploit new research areas.

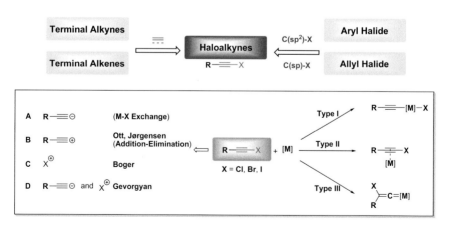

Scheme 1.1 Potential reaction pathways of haloalkynes in transition metal catalysis

References

1. Li C, Trost BM (2008) Green chemistry for chemical synthesis. Proc Natl Acad Sci 105:13197–13202
2. Wu W, Jiang H (2014) Haloalkynes: a powerful and versatile building block in organic synthesis. Acc Chem Res 47:2483–2504
3. Laurence C, Queignec-Cabanetos M, Dziembowska T, Queignec R, Wojtkowiak B (1981) 1-Iodoacetylenes. 1. Spectroscopic evidence of their complexes with Lewis bases. A spectroscopic scale of soft basicity. J Am Chem Soc 103:2567–2573
4. Gao K, Goroff NS (2000) Two new iodine-capped carbon rods. J Am Chem Soc 122: 9320–9321
5. Rege PD, Malkina OL, Goroff NS (2001) The effect of Lewis bases on the ^{13}C NMR of Iodoalkynes. J Am Chem Soc 124:370–371
6. Ott E, Dittus G (1943) Polymerisation und Reaktionen mit Ammoniak, Aminen, Alkoholaten und Natriummalonester. Chem Ber 76:80–84
7. Poulsen TB, Barnardi L, Aleman J, Overgaard J, Jørgensen KA (2007) Organocatalytic asymmetric direct α-alkynylation of cyclic β-ketoesters. J Am Chem Soc 129:441–449
8. Boger DL, Brunette SR, Garbaccio RM, Boger DL, Brunette SR, Garbaccio RM (2001) Synthesis and evaluation of a series of C3-substituted CBI analogues of CC-1065 and the Duocarmycins. J Org Chem 66:5163–5173
9. Trofimov A, Chernyak N, Gevorgyan V (2008) Dual role of Alkynyl halides in one-step synthesis of Alkynyl epoxides. J Am Chem Soc 130:13538–13539
10. Brand JP, Waser J (2012) Electrophilic alkynylation: the dark side of acetylene chemistry. Chem Soc Rev 41:4165–4179

Chapter 2
Preparation of Haloalkynes

Abstract Traditionally, haloalkynes were accessible through the deprotonation of the corresponding terminal alkynes with a strong base, followed by trapping with a halogenating reagent. During the past decades, several mild and convenient methods have been developed, thus increasing the attractiveness of this class of compounds in organic synthesis. Among which, the electrophilic bromination of terminal alkynes with *N*-bromosuccinimide (NBS) and Ag catalyst is one of the most commonly used methods for the preparation of bromoalkynes due to the mild reaction conditions, high efficiency and simple manipulation. In this chapter, we will detailedly describe the general and practical methods to prepare bromoalkynes, chloroalkynes, and iodoalkynes.

Keywords Terminal alkynes · Ag catalyst · Bromoalkyne synthesis · Chloroalkyne synthesis · Iodoalkyne synthesis

Haloalkynes were traditionally accessible through the deprotonation of the corresponding terminal alkynes with a strong base, followed by trapping with a halogenating reagent. Recently, several mild and convenient methods have been developed (Scheme 2.1), [1] thus increasing the attractiveness of this class of compounds in organic synthesis. Among which, the electrophilic bromination of terminal alkynes with *N*-bromosuccinimide (NBS) and Ag catalyst [2] is one of the most commonly used methods for the preparation of bromoalkynes due to the mild reaction conditions, high efficiency and simple manipulation (Scheme 2.2) [3].

General Procedure for the Synthesis of Bromoalkynes: To a solution of alkyne (1 equiv) in acetone (0.2 mmol/mL) was added NBS (1.1 equiv) and AgNO$_3$ (10 mol%) at room temperature with magnetic stirring. After 2–3 h, the reaction mixture was diluted with hexanes (100 mL) and filtered off the crystals formed. The filtrate was concentrated under reduced pressure and passed through a pad of silica gel using hexanes as an eluent. The filtrate was collected and evaporated under reduced pressure to afford a pure colorless oil of bromoalkyne.

General Procedure for the Synthesis of Chloroalkynes: To a solution of alkyne (1 equiv) in CCl$_4$ (2 mmol/mL) was added Cs$_2$CO$_3$ (1.1 equiv) and Bu$_4$NCl (5 mol %) at 70 °C with magnetic stirring. After 6–7 h, the reaction mixture was diluted

H. Jiang et al., *Haloalkyne Chemistry*, SpringerBriefs in Green Chemistry for Sustainability, DOI 10.1007/978-3-662-49001-3_2

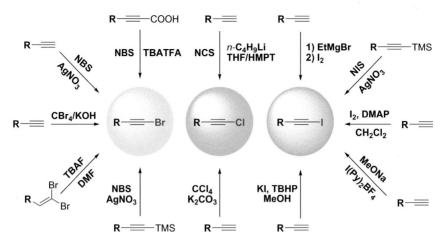

Scheme 2.1 Preparation methods for haloalkynes

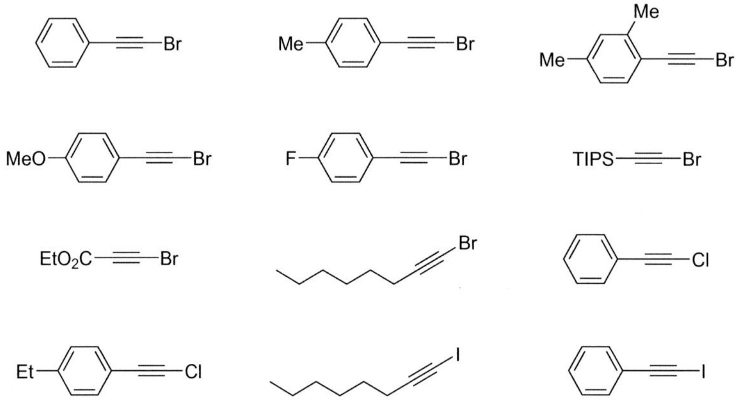

Scheme 2.2 Representative haloalkynes

with hexanes (100 mL) and filtered off the crystals formed. The filtrate was concentrated under reduced pressure and passed through a pad of silica gel using hexanes as an eluent. The filtrate was collected and evaporated under reduced pressure to afford a pure colorless oil of chloroalkyne.

General Procedure for the Synthesis of Iodoalkynes: To a solution of alkyne (1 equiv) in acetone (0.2 mmol/mL) was added NIS (1.1 equiv) and AgNO₃ (10 mol %) at room temperature with magnetic stirring. After 2–3 h, the reaction mixture was diluted with hexanes (100 mL) and filtered off the crystals formed. The filtrate was concentrated under reduced pressure and passed through a pad of silica gel using hexanes as an eluent. The filtrate was collected and evaporated under reduced pressure to afford a pure colorless oil of iodoalkyne.

References

1. For a review on preparation of haloalkynes, see: Zhdankin VV (1995) Alkynyl halides and chalcogenides. In: Katritzky AR, Meth-Cohn O, Rees CW (eds) Comprehensive organic functional group transformations. Pergamon, New York, pp 1011–1038
2. Nishikawa T, Shibuya S, Hosokawa S (1994) One pot synthesis of haloacetylenes from trimethylsilylacetylenes. Synlett 7:485–486
3. Gao Y, Yin M, Wu W, Huang L, Jiang H (2013) Copper-catalyzed intermolecular oxidative cyclization of haloalkynes: synthesis of 2-halo-substituted imidazo[1,2-a]pyridines, Imidazo [1,2-a]pyrazines and Imidazo[1,2-a]pyrimidines. Adv Synth Catal 355:2263–2273

Chapter 3
Reactions of Haloalkynes

Abstract Haloalkynes are a significant class of molecules that have been widely utilized in organic synthesis. In this chapter, we will describe representative examples of haloalkynes, with particular attention paid to the reaction design and mechanistic investigation as well as the general experimental procedures. According to the reactive sites of haloalkynes involved in the transformations, the reactions are classified to three types: (i) the transformations of carbon-halo bond motif; (ii) the diverse functionalization of carbon-carbon triple bond unit; (iii) the reactions involved both carbon-halo bond motif and carbon-carbon triple bond unit. These transformations present a powerful tool for haloalkynes to construct molecular complexity efficiently.

Keywords Haloakyne reagents · Diverse transformations · Reactive sites · Reaction design · Mechanism investigation

In this chapter, the emphasis will be put on the reaction development of haloalkynes. According to the reactive sites of haloalkynes involved in the transformations, the reactions are classified to three types: (i) the transformations of carbon-halo bond motif (Scheme 3.1, path A); (ii) the diverse functionalization of carbon-carbon triple bond unit (Scheme 3.1, path B); (iii) the reactions involved both carbon-halo bond motif and carbon-carbon triple bond unit (Scheme 3.1, path C). These transformations present a powerful tool to construct molecular complexity efficiently. Representative examples are described, with particular attention paid to the reaction design and mechanistic investigation.

3.1 Transformations of Carbon-Halo Bond Motif

Alkyne motif is one of the most important and useful building blocks in natural products, pharmaceuticals, as well as functional materials. Subsequently, the construction of alkyne motif contained molecules has attracted considerable attention during the past decades. Transformations of haloalkyne reagents based on the

© The Author(s) 2016

H. Jiang et al., *Haloalkyne Chemistry*, SpringerBriefs in Green Chemistry
for Sustainability, DOI 10.1007/978-3-662-49001-3_3

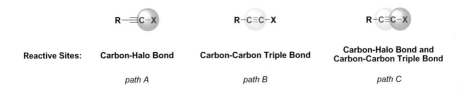

Reactive Sites: Carbon-Halo Bond   Carbon-Carbon Triple Bond   Carbon-Halo Bond and
                                                                              Carbon-Carbon Triple Bond

                    path A   *path B*   *path C*

Scheme 3.1 Reactive sites of haloalkynes

highly reactive carbon-halo bond, could realize the facile synthesis of skeletons that previously were unable or difficult to prepare.

3.1.1 Construction of Carbon-Carbon Bond

3.1.1.1 Construction of C(sp)–C(sp) Bond

Conjugated diynes, especially 1,3-diyne compounds, are of vital significance as versatile building blocks in the synthesis of natural products, bioactive compounds, as well as functional materials [1, 2]. Consequently, the construction of conjugated diynes has attracted great attention for a long time.

C(sp)–C(sp) Homo-Coupling

Due to the importance of symmetrical 1,3-diyne compounds, they are usually synthesized either by Cu-catalyzed homo-coupling reactions including Glaser coupling [3], Eglinton coupling [4], Hay coupling [5, 6], and Pd-mediated homo-coupling reactions [7] or other related modified methods [8]. In this context, the homo-coupling of haloalkynes would provide an alternative route to access 1,3-diyne compounds. In 2003, Lee's group [9] documented a highly efficient Pd^0-catalyzed homo-coupling reaction of 1-iodoalkynes to construct symmetrical 1,3-diynes under mild and simple reaction conditions. This reaction did not use copper salts or other metal reagents and a base, and was conducted under an inert atmosphere, thus preventing side reactions associated with the Glaser coupling reaction in which O_2 is usually used as the oxidant. Although the exact mechanism of this homo-coupling reaction of 1-iodoalkynes was not clear yet, the author proposed that the formation of dialkynylpalladium intermediate (**1**) through the oxidative addition product of Pd^0 to 1-iodoalkyne reacted with another 1-iodoalkyne, which then converted to the 1,3-diyne and iodine (Scheme 3.2).

General Procedure for Pd-Catalyzed Homo-Coupling Reactions of Iodoalkynes: To Pd(PPh$_3$)$_4$ (4 mol%) was added a solution of 1-iodoalkyne (1 mmol) in dry *N,N*-dimethylformamide (DMF) (2 mL) under a nitrogen atmosphere. After 2–5 h, the mixture was poured into an aqueous saturated NaHCO$_3$ solution (15 mL) and then extracted with diethyl ether (15 mL × 3). The combined organics were washed with

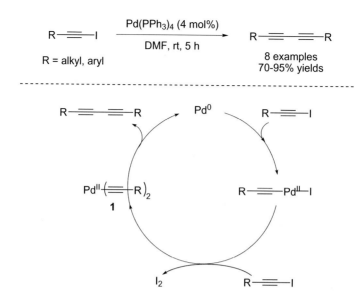

Scheme 3.2 Pd-catalyzed homo-coupling reactions of iodoalkynes

brine (15 mL), dried with anhydrous MgSO$_4$, filtered, and concentrated under reduced pressure. The residue was purified by silica gel column chromatography to give the corresponding 1,3-diyne.

Given the environmental and economical factors, the development of transition metal-free reaction systems for the construction of 1,3-diynes is highly demanded. In 2010, Jiang's group [10] successfully developed an efficient synthetic method to 1,3-diynes from haloalkynes under the treatment of KI in DMF solvent. This approach also featured both oxidant and base free (Scheme 3.3). Generally, better yields of symmetrical 1,3-diynes were obtained from iodoalkynes than the corresponding bromoalkynes. Both aromatic and aliphatic alkynyl halides could perform this homo-coupling reaction smoothly under the standard reaction conditions. And diverse functional groups on haloalkyne substrates, such as fluoro, chloro, hydroxyl, nitrile group could be tolerated. To the reaction mechanism, the author proposed the involvement of iodoalkyne intermediate generated from the substitution of bromoalkyne with KI, which might undergo two pathways for the obtained 1,3-diyne product. The iodoalkyne would be transformed to iodine- and alkyne-radicals, which were then homo-coupled to deliver the symmetrical 1,3-diyne and iodine (Scheme 3.3, path A). Alternatively, the iodoalkyne intermediate was decomposed to iodine and alkyne anion, followed by a redox process to give the alkyne radical, which would undergo further transformation to afford the final product (Scheme 3.3, path B).

General Procedure for KI-Mediated Homo-Coupling Reactions of Haloalkynes: Haloalkyne (1 mmol) and KI (3 mmol) in DMF (2 mL) were stirred at 120 °C for 12 h in a Schlenk tube (25 mL). Water (8 mL) was added after the completion of the

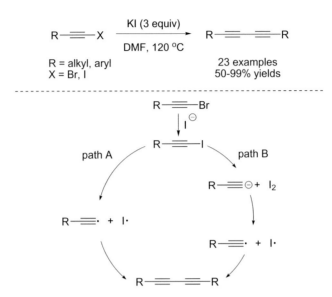

Scheme 3.3 KI-mediated homo-coupling reactions of haloalkynes

reaction, the aqueous solution was extracted with diethyl ether (15 mL × 3), and the combined extracts were dried with anhydrous $MgSO_4$, filtered, and concentrated under reduced pressure. The residue was purified by silica gel column chromatography to give the corresponding 1,3-diyne.

C(sp)–C(sp) Cross-Coupling

Compared to symmetrical 1,3-diyne compounds, the synthesis of unsymmetrical 1,3-diynes is rather difficult, due to the competition of homo-coupling reaction. Among the methods for the synthesis of unsymmetrical 1,3-diynes, Cadiot-Chodkiewicz cross-coupling reaction is one of the most representative examples [11]. In 2007, Jiang's group documented a mild and environmentally friendly method for the Cu-catalyzed Cadiot-Chodkiewicz coupling of bromoalkynols with terminal acetylenes in $scCO_2$ utilizing NaOAc as base (Table 3.1) [12]. Methanol, as a co-solvent, could improve the dissolution of inorganic salts in $scCO_2$ and facilitate the reaction rate. This new cross-coupling reaction system not only tolerated a wide range of functional groups to deliver diverse unsymmetrically substituted 1,3-diynes, but also avoided the employment of amine. However, experiment results revealed that this transformation was sensitive to the pressure of $scCO_2$ and high reaction temperature.

    General   Procedure   for   Cu-Catalyzed   Cross-Coupling   Reactions   of Bromoalkynols: CuCl (5 mol%), AcONa (1.5 mmol), MeOH (1 mL), bromoalkyne (1 mmol) and alkyne (1.2 mmol) were added to an autoclave vessel (15 mL) in

Table 3.1 Cu-catalyzed cross-coupling reactions of bromoalkynols

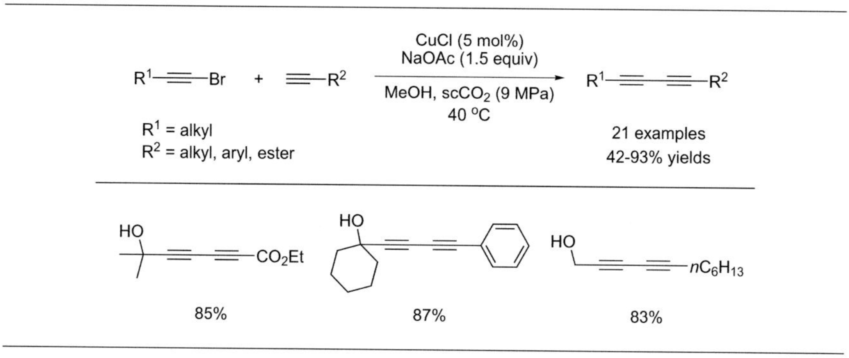

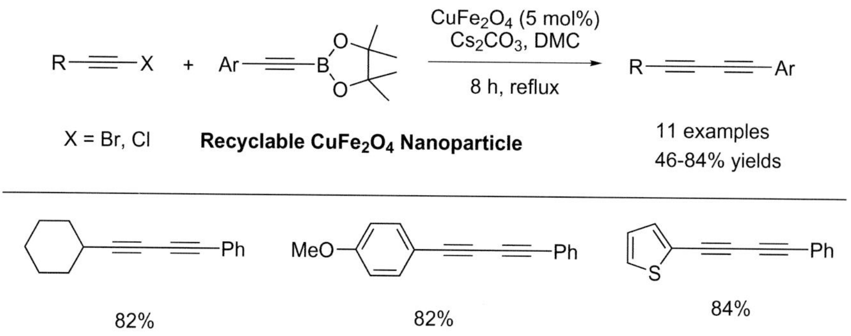

85%   87%   83%

X = Br, Cl   **Recyclable CuFe$_2$O$_4$ Nanoparticle**   11 examples
                                                                          46-84% yields

82%   82%   84%

Scheme 3.4 CuFe$_2$O$_4$ nanoparticle catalyzed cross-coupling reactions of haloalkynes

sequence. Liquid CO$_2$ was pumped into the autoclave by a cooling pump until the desired pressure was reached then the autoclave was heated in an oil bath under magnetic stirring for the desired reaction time. After the reaction was completed, the autoclave was allowed to cool to 0 °C and CO$_2$ was vented. The residue was extracted with Et$_2$O (20 mL). The extract was filtered and concentrated under reduced pressure to give a residue that was purified by chromatography on a silica gel column using light PE-EtOAc as eluent.

Later in 2014, Ranu and co-workers developed a novel protocol for C(sp)–C(sp) cross-coupling of haloalkynes with pinacol ester of alkynyl boronic acid in dimethyl carbonate (DMC) using a commercially available and magnetically separable CuFe$_2$O$_4$ nanoparticle catalyst. The reaction has a broad substrate scope and tolerates diverse functional groups. Importantly, the CuFe$_2$O$_4$ nanoparticle catalyst was recycled more than 10 times with marginal loss of activity in subsequent runs (Scheme 3.4) [13].

General Procedure for CuFe$_2$O$_4$ Nanoparticle Catalyzed Cross-Coupling Reactions of Haloalkynes: A suspension of haloalkyne (1 mmol), pinacol ester of

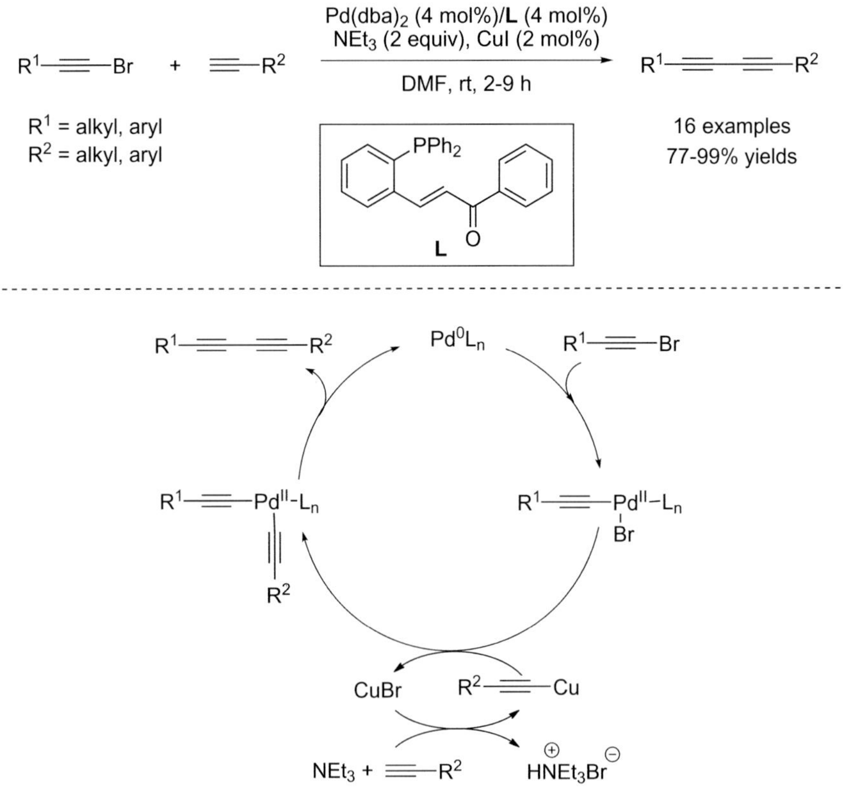

Scheme 3.5 Ligand accelerated Pd-catalyzed cross-coupling reactions of bromoalkynes

alkynelboronic acid (1.5 mmol), Cs_2CO_3 (2 mmol), and $CuFe_2O_4$ (5 mol%) in DMC (5 mL) was stirred at 100 °C (oil bath temperature) for 8 h under argon. The reaction mixture was allowed to cool and extracted with ethyl acetate (20 mL × 3). The organic extracts were washed with brine, dried with anhydrous Na_2SO_4, filtered, and concentrated under reduced pressure. The product was obtained by flash column chromatography.

Except for Cu catalysis, Pd complexes also serve as a powerful catalysts to construct several conjugated diynes [14]. However, the competitive homo-coupling process is still the major challenge in Pd-catalyzed C(sp)–C(sp) cross-coupling reactions. In 2008, Lei's group [15] reported an efficient method to synthesize unsymmetrical 1,3-diynes which was promoted by Pd(dba)$_2$ with a phosphine-olefin ligand **L** (Scheme 3.5). This protocol realized the cross-coupling reaction of a wide spectrum of terminal alkynes and haloalkynes, affording the corresponding conjugated diynes in good to excellent yields with high selectivity. Notably, one-pot synthesis of symmetrical and unsymmetrical triynes was also achieved. Mechanistic investigations indicated that the phosphine-olefin ligand could accelerate the reductive elimination process in the catalytic cycle, thereby enhancing the selectivity.

General Procedure for Ligand Accelerated Pd-Catalyzed Cross-Coupling Reactions of Bromoalkynes: To an oven-dried Schlenk tube with a magnetic stir bar were added Pd(dba)$_2$ (4 mol%), **L** ligand (4 mol%), and CuI (2 mol%). DMF (1 mL) was added via a syringe. The system was vacuumed with an oil pump at 0 °C and filled with nitrogen, and this procedure was repeated five times. After the mixture was stirred under nitrogen for about 10 min, alkyne (0.6 mmol) was added via a microliter and stirred for another 5 min. 1-Bromoalkyne (0.5 mmol) was added last via a microliter syringe. The system was stirred at room temperature for 10 h. Upon completion, brine (4 mL) was added, and the mixture was extracted by ethyl acetate (3 mL × 3), and the combined extracts were dried with anhydrous MgSO$_4$, filtered, and concentrated under reduced pressure. The product was obtained by flash column chromatography.

Later in 2012, Lei and co-workers developed a more efficient Pd^0-catalyzed C(sp)–C(sp) cross-coupling reaction of terminal alkynes with bromoalkynes. Interestingly, the reaction could run at 100 mmol scale, and more than 99 % of the cross-coupling product was obtained without any bromoalkyne homo-coupled by-product. The key to the success of this transformation was the utilization of TBAB (tetrabutylammonium bromide) as a stabilizer, which could prevent the aggregation and precipitation of palladium catalyst. Even when the catalyst loading was reduced to 0.01 mol%, the reaction could still proceed efficiently, and the catalyst was kept active. Kinetic studies indicated that the reaction rate was not first order to Pd catalyst via in situ IR spectroscopy, and only part of the Pd species was employed to catalyze this C(sp)–C(sp) cross-coupling reaction. Importantly, palladium nanoparticles were observed in this reaction (Scheme 3.6) [16].

General Procedure for TBAB Stabilized Pd-Catalyzed Cross-Coupling Reactions of Bromoalkynes: A mixture of haloalkyne (1 mmol), terminal alkyne (1.5 mmol), TBAB (0.3 mol%), and CuI (0.2 mol%) in iPr$_2$NH (5 mL) was stirred under N$_2$ at 70 °C for 5 min. Then Pd(OAc)$_2$ (0.01 mol%) was added in one

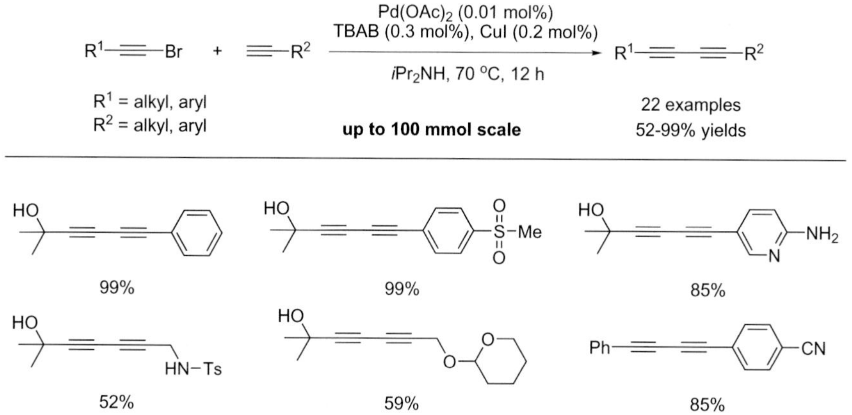

Scheme 3.6 TBAB stabilized Pd-catalyzed cross-coupling reactions of bromoalkynes

portion. After reaction completion, as indicated by TLC and GC, the mixture was quenched with diluted hydrochloric acid (4 mL, 2 M), and the solution was extracted with ethyl acetate (15 mL × 3). The organic layers were combined and dried over sodium sulfate. The pure product was obtained by flash column chromatography on silica gel.

3.1.1.2   Construction of C(sp)–C(sp²) Bond

Functionalized cyclic and acyclic enynes are all-pervading subunits in a wide range of natural products, functional materials and bioactive compounds [17]. To this regard, the development of efficient and practical methods for the construction of conjugated enyne molecules has become the subject of intensive investigation in the area of synthetic and medicinal chemistry.

C(sp)–C(sp²) Cross-Coupling

In 1985, Suzuki and co-workers reported the synthesis of alkyenynes via palladium-catalyzed cross-coupling reaction of 1-alkenylboranes with bromoalkynes (Scheme 3.7) [18]. This reaction was stereo- and regiospecifically, and the

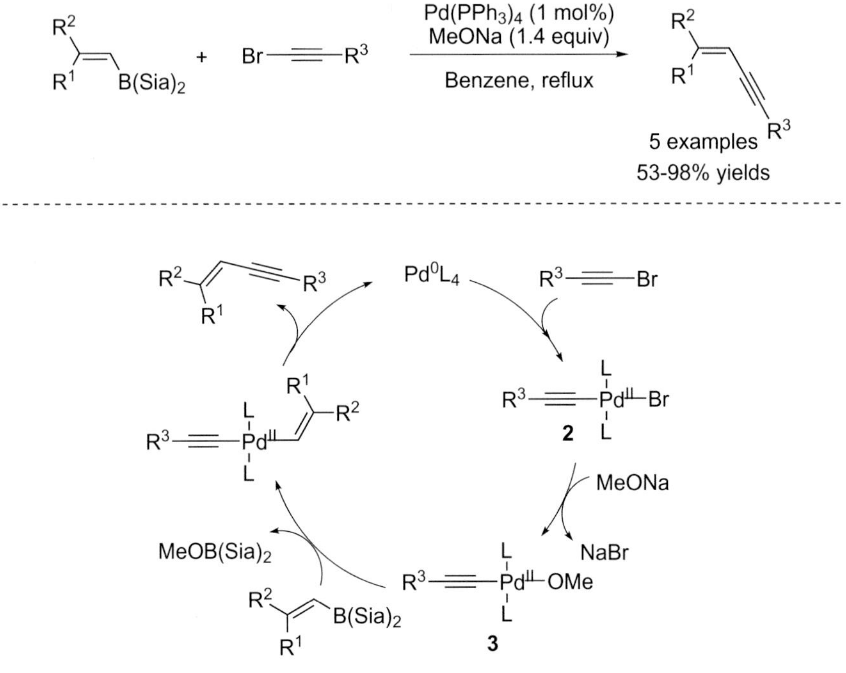

Scheme 3.7 Cross-coupling reaction of 1-alkenylboranes with bromoalkynes

configurations of both the starting alkynylboranes and bromoalkynes were retained. Mechanistic studies indicated that the transmetalation between an alkynylborane and an alkoxypalladium(II) complex **3** generated through the metathetical displacement of a halogen atom from the intermediate **2** with the base. Later, the cross-coupling reactions of haloalkynes with activated alkenes have been broadly investigated, such as alkenylboronic acid [19–21]. oragnozinc reagents [22, 23], vinylstannanes [24–26]. Grignard reagents [27], vinylzirconocene [28, 29], and vinylsiloxanes [30].

General Procedure for Palladium Catalyzed Cross-Coupling Reaction of 1-Alkenylboranes with Bromoalkynes: A flask (50 mL) was charged with Pd(PPh$_3$)$_4$ (1 mol%), dry benzene (12 mL), and bromoalkyne (5 mmol) under a nitrogen atmosphere. The reaction mixture was stirred for 30 min at room temperature, and to the solution were added alkenylborane (6 mmol) and MeONa (7 mmol, 1 M in MeOH). The reaction mixture was heated under reflux for 2 h and then treated with aqueous NaOH (1.8 mL, 3 M solution) and H$_2$O$_2$ (1.8 mL of a 30 % solution) for 1.5 h at room temperature to remove the unreacted alkenylborane. The product was extracted with hexane and dried over MgSO$_4$. After the removal of the solvent, the enyne product was purified by distillation.

Alternatively, the Sonogashira coupling, the cross-coupling reaction between vinyl halides and terminal alkynes also represents one of the most widely used strategies achieving the synthesis of functionalized enynes [31]. In this context, the development of "Inverse Sonogashira Coupling", the term which was first introduced by Trofimov [32], has attracted more and more attention (Scheme 3.8) [33]. It stands as a complementary strategy for the synthesis of aryl/heteroaryl alkynes through the direct alkynylation of unreactive C(sp^2)–H bonds with readily available haloalkynes. In 1992, Kalinin and coworkers firstly reported this type of alkynylation reaction with a stoichiometric amount of Cu^I salt [34]. Later, Trofimov and co-workers devoted great efforts in this area [35–42]. However, the major breakthrough was achieved until 2007, as the first example of a transition metal-catalyzed direct alkynylation of electron-rich *N*-fused heterocycles was promulgated by Gevorgyan's group (Scheme 3.9) [43]. A wide spectrum of indolizine, pyrroloisoquinoline, pyrroloquinoline, and pyrrolooxazole derivatives could be regioselectively alkynylated with different substituted bromoalkynes in the presence

Scheme 3.8 Development of "Inverse Sonogashira Coupling"

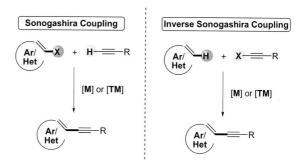

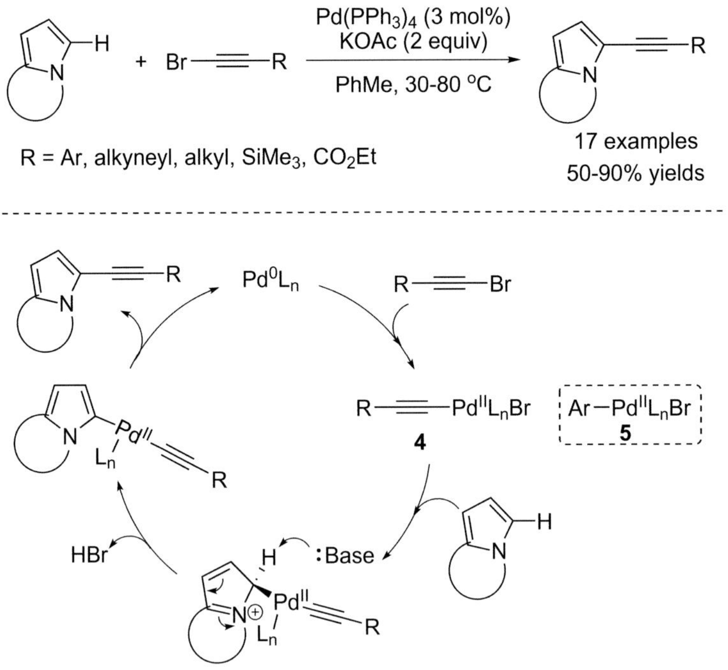

Scheme 3.9 Pd-catalyzed alkynylation of *N*-fused heterocycles

of Pd catalyst. The alkynylpalladium intermediate **4**, which is essential for conceptual advance, generated through Pd^0 catalyst oxidative addition into the C–Br bond of bromoalkyne, exhibited the similar reactivity to that of arylpalladium species **5**, which is well known to undergo an electrophilic pathway in the process of indolizine arylation. Subsequently, Gu [44], Chang [45], Jiang [46, 47], and Loh [48] documented different cross-coupling partners with haloalkynes under palladium catalysis.

General Procedure for Pd-Catalyzed Alkynylation of N-Fused Heterocycles: In a glovebox under nitrogen atmosphere, to a Wheaton microreactor (5 mL) equipped with a spin vane and screw cap with a polytetrafluoroethylene (PTFE) faced silicone septum under nitrogen atmosphere were added heterocyclic substrate (1 equiv), Pd(PPh$_3$)$_2$Cl$_2$ (3–5 mol%) and KOAc (2 equiv). The microreactor was removed from the glovebox, bromoalkyne (1.3–1.8 equiv) and anhydrous toluene (0.001–0.010 M) were successively added and the mixture was stirred until completion (as monitored by TLC and/or GC/MS). The solvent was removed under reduced pressure and the residue was purified using flash-column chromatography using hexane or hexanes/ethylacetate combination as eluent to afford pure alkynyl-heterocycles.

Despite palladium catalysis, Piguel [49] reported a more efficient general procedure for the direct alkynylation of various heterocycles with copper-catalysis. The author found out that the success of the copper bromide/dimethyl sulfide complex

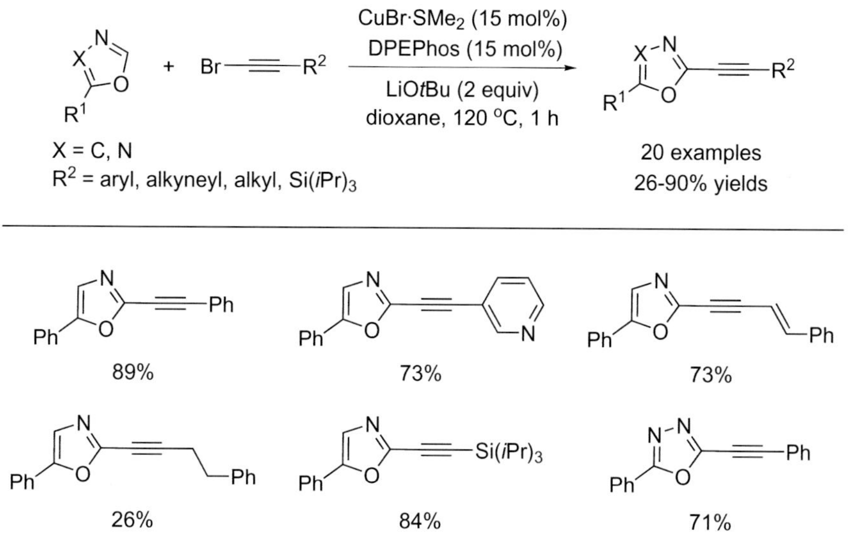

Scheme 3.10 Copper-catalyzed alkynylation of azoles

lied in the better solubility compared that with the uncomplexed copper bromide, which allowed it to coordinate immediately in the reaction medium. This method did not need high dilution reaction condition. Especially, the minimal cost and toxicity of copper catalyst could tolerate various oxazoles with different electron property and structural diversity, and gave the coupling product in good yields (Scheme 3.10). Notably, the azoles are with high pK_a values (>30), while the carbon-halo bond of the substrate could survive under the standard reaction conditions. Additionally, the structure of the product was unambiguously confirmed by single-crystal X-ray diffraction analysis.

General Procedure for Copper-Catalyzed Alkynylation of Azoles: A flame dried tube under argon was charged with azole (0.69 mmol), CuBr·SMe$_2$ (15 mol%), DPE-Phos (15 mol%), *t*BuLi (1.38 mmol). Then bromoalkyne (1.38 mmol) was diluted into dioxane (2 mL) and the solution was added to the medium. The tube was sealed with Teflon cap and put in a pre-heated oil bath at 120 °C for 1 h. The reaction mixture was diluted with ethyl acetate and water was added. This mixture was extracted with ethyl acetate and the combined organic layers were put together and dried over MgSO$_4$. Solvents were removed under reduced pressure and the crude was purified by flash chromatography on silica gel to afford the desired product.

Based on the experimental results and previous literatures [50, 51], the authors proposed the possible reaction mechanism as illustrated in Scheme 3.11. Firstly, the deprotonation of oxazole lithium base, followed by lithium-copper transmetallation to generate Cu^I intermediate **6**. Subsequently, the oxidative addition of **6** to bromoalkyne gave the four-coordinated Cu^{III} complex **7**. Finally, the reductive elimination led to the desired alkynylated product, and regenerated the catalytic Cu^I

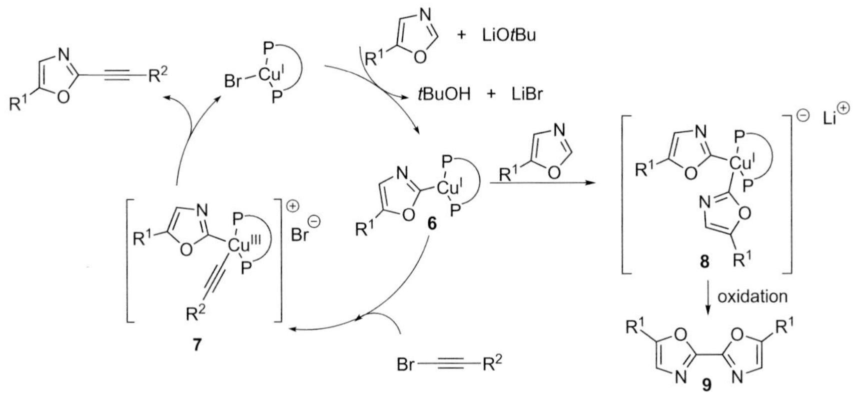

Scheme 3.11 Proposed reaction mechanism

species. Undoubtedly, the formation of Cu^{III} complex **7** would compete with activation of another oxaole, which would deliver the bis(oxazolte) Cu^{I} species **8**, and afforded the undersired bis(oxazole) dimer **9**. Importantly, the steric hindrance of the copper center favored the less sterically demanding haloalkyne, thus **7** was formed preferentially. Finally, due to the solubility of dimethylsulfide complex and the bulky steric hindrance effect of the ligand, the catalytic cycle was driven towards the formation of the alkynylated product.

Later, Miura [52] and Das [53] independently realized the copper-catalyzed alkynylation of 1,3,4-oxadiazoles. In the mean time, other transition metals, such as nickel, were found to be suitable catalysts for the inverse Sonogashira coupling reactions of haloalkynes [54]. These achievements in the area of direct alkynylation reactions involving haloalkyne reagents open up new exciting opportunities for the functionalization of diverse C(sp^2)–H bonds.

The cyclization of alkynes bearing proximate nucleophilic centers promoted by organopalladium complexes is an effective strategy for heterocyclic ring synthesis [55]. This chemistry provides a direct method to the construction of functionalized cycles through the regio- and stereoselective addition of a nucleophile to the carbon-carbon triple bond with the generation of a vinylpalladium complex, which could proceed diverse transformations (Scheme 3.12). Taking the advantages of this strategy, Larock [56], Cacchi [57], and Yorimitsu [58] independently demonstrated that nucleophilic addition triggered cross-coupling reaction of haloalkynes to synthesize 3,4-disubstituted isoquinolines, 3-alkynylindoles and 1,2-disubstituted cyclopentenes, respectively.

Alkynylation Reactions

Compared to cross coupling reactions, electrophilic ethynylation of carbon nucleophiles is another attractive methodology to construct C(sp)–C(sp^2) bonds. In 2002,

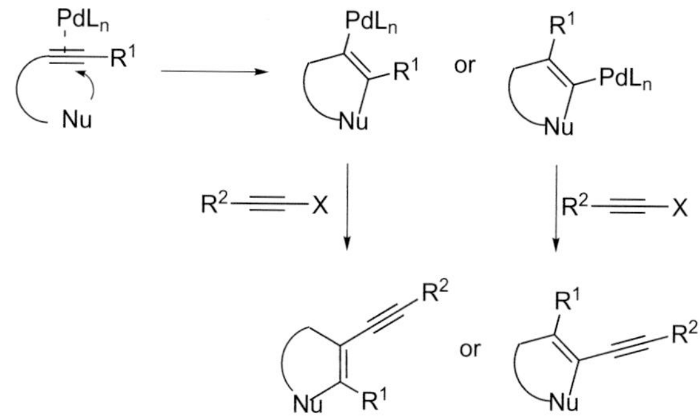

Scheme 3.12 Pd-catalyzed nucleophilic addition triggered cross-coupling reaction

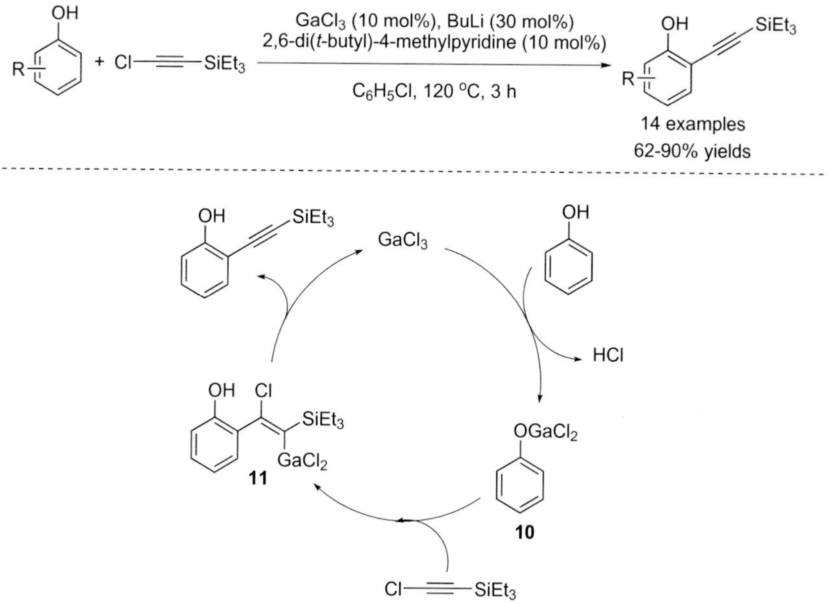

Scheme 3.13 GaCl$_3$-catalyzed *ortho*-ethynylation of phenols

Yamaguchi and co-workers reported GaCl$_3$-catalyzed ortho-ethynylation of phenols. Various substituted phenols were applicable to this method, and the turnover number based on the catalyst (GaCl$_3$) was between 8 and 10. The mechanism studies indicated this catalytic ethynylation involved carbogallation of haloalkyne and the formation of intermediate **10** under the effect of lithium salts. Interestingly, the protonated product of intermediate **11** was not detected in the reaction mixture. It seemed that β-elimination of **11** happened to be more rapid (Scheme 3.13) [59]. The author also applied this method to the *ortho*-ethynylation of silyl enol ethers

[60] and anilines [61]. Later, Chatani [62, 63], and Chen [64] independently reported palladium-catalyzed *ortho*-alkynylation of $C(sp^2)$–H bond in benzenes with different directing groups.

General Procedure for GaCl₃-Catalyzed ortho-Ethynylation of Phenols: Under an argon atmosphere, to a solution of phenol (10 mmol) in chlorobenzene (50 mL) were added butyllithium (3 mmol, 1.6 M in hexane) and $GaCl_3$ (1 mmol, 1 M in methylcyclohexane) at 0 °C successively. The mixture was stirred for 10 min at room temperature, and then 2,6-di(*tert*-butyl)-4-methylpyridine (1 mmol) and chlorotriethylsilylethyne (10 mmol) were added. The mixture was heated at 120 °C for 3 h. Water (25 mL) and THF (25 mL) were added, and the organic materials were extracted with ethyl acetate, washed with water and brine, dried over $MgSO_4$, filtered, and concentrated in vacuo. The residue was purified by flash chromatography over silica gel to provide the pure product.

On the other hand, the development of practical and efficient alkynylation methods for functionalized acyclic enyne compounds is also highly demanded. Undoubtedly, one of the most straightforward strategies to achieve this goal is the direct addition of an "activated" alkyne to another alkyne. In this context, several catalytic systems for alkynylstannylation [65], alkynylzirconation [66, 67], alkynylboranation [68], and alkynylcyanation [69] have been developed (Scheme 3.14a). In 2010, Jiang and coworkers revealed a Pd-catalyzed selective intermolecular cross-coupling reaction between haloalkynes and internal alkynes, delivering various halogenated enyne products through a new type of direct bromoalkynylation process (Scheme 3.14b) [70]. Condition optimization indicated that Pd^{II} was crucial to the product formation, while Pd^0 just impeded the reaction. Reductive additives and inorganic bases would also retard the transformation. However, air or organic oxidant did not interrupt the reaction. This approach was found to have a broad substrate scope (Table 3.2). A wide range of haloalkynes, including aryl-, alkynyl-, and trimethylsilyl- alkynyl bromides, were able to proceed

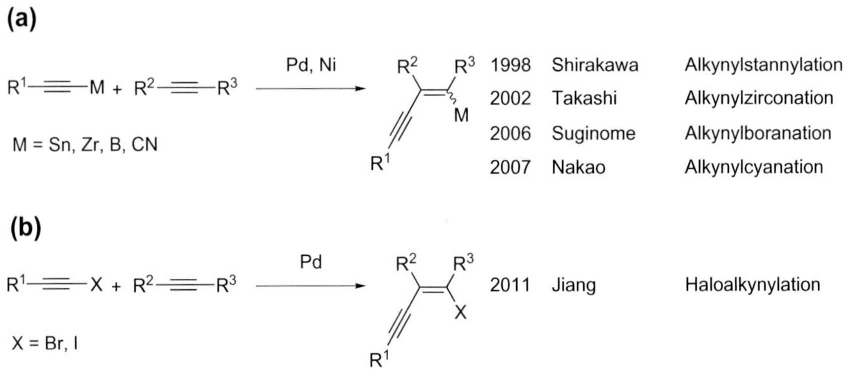

(a)

	Pd, Ni				
R¹—≡—M + R²—≡—R³ →			1998	Shirakawa	Alkynylstannylation
			2002	Takashi	Alkynylzirconation
M = Sn, Zr, B, CN			2006	Suginome	Alkynylboranation
			2007	Nakao	Alkynylcyanation

(b)

	Pd				
R¹—≡—X + R²—≡—R³ →			2011	Jiang	Haloalkynylation
X = Br, I					

Scheme 3.14 Strategies for the conjugated enyne synthesis. **a** Previous alkynylation strategies. **b** Jiang's strategy

Table 3.2 Pd-catalyzed bromoalkynylation of alkynes

this bromoalkynylation reaction smoothly to afford the corresponding products in good to excellent yields. Additionally, reasonable yields were achieved when this reaction was extended to iodoalkynes instead of bromoalkynes. Importantly, exclusively *cis*-addition products were obtained for symmetrical internal alkynes. While the regioselectivity of the unsymmetrical disubstituted acetylenes was mainly influenced by the functional groups in the internal alkynes.

General Procedure for Pd-Catalyzed Bromoalkynylation of Alkynes: To a Schlenk tube (25 mL) was successively added Pd(OAc)$_2$ (5 mol%), CH$_3$CN (2 mL), 4-octyne (1 mmol) and haloalkyne (1.2 mmol). The resulting mixture was stirred at 30 °C for 8 h. Then, the mixture was filtered through a small amount of silica gel. The filtrate was concentrated under reduced pressure, and the residue was purified by silica gel preparative TLC (*n*-hexane) to give the desired product.

To gain some insight of the reaction mechanism, the authors performed some control experiments with stoichiometric Pd catalysts and the major halogenated products were identified to be originated from phenylethynyl halides [Scheme 3.15, Eqs. (1) and (2)]. These results provided evidence of a mechanism that Pd^{II} species underwent an unusual oxidative addition to phenylethynyl bromide, rather than a direct halopalladation reaction of alkynes. Accordingly, the mechanism of this transformation was initiated by the oxidative addition of Pd^{II} salt to bromoalkyne to form the Pd^{IV} complex **12**. Then, the *cis*-addition of **12** to internal alkyne afforded the *cis*-alkynyl vinylpalladium intermediate **13**, which underwent a reductive elimination to deliver the brominated enyne product and regenerate the active Pd^{II} catalyst (Scheme 3.15).

Yn-1-imines are an important class of compounds with novel π-system, which have wide applications in functional materials [71]. In 2011, an elegant three-component coupling reaction of arynes, isocyanides and bromoalkynes for the

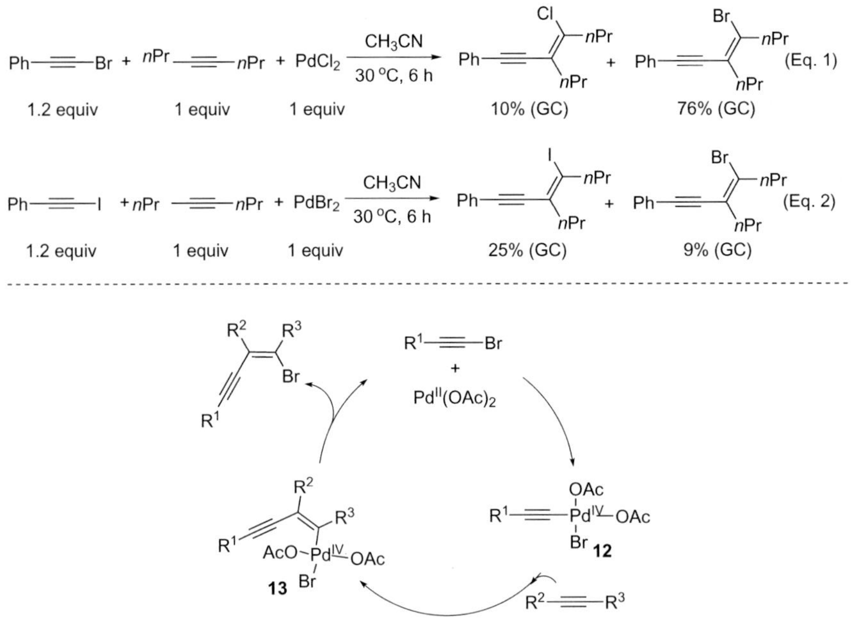

Scheme 3.15 Control experiments and proposed catalytic cycle

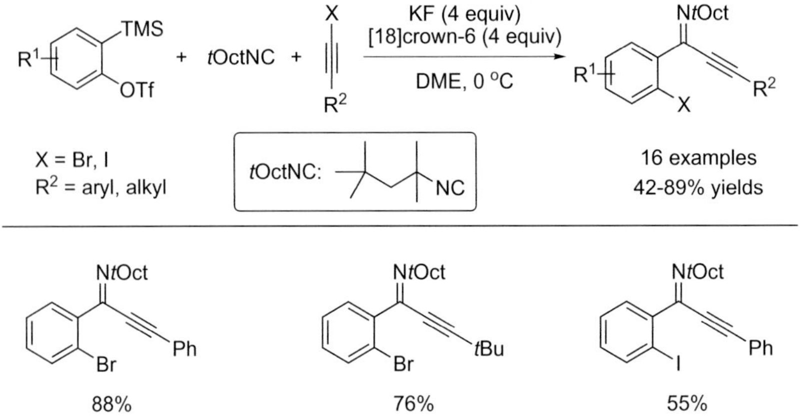

Scheme 3.16 Three-component coupling of arynes, isocyanide and haloalkynes

synthesis of yn-1-imine compounds has been reported by Yoshida and co-workers [72]. The benzyne was in situ generated from its precursor under the treatment of KF/[18] crown-6. Importantly, this reaction adapted to broad substrate scope and tolerated various functional groups (Scheme 3.16).

General Procedure for Three-Component Coupling of Arynes, Isocyanide and Haloalkynes: A Schlenk tube equipped with a magnetic stirring bar was charged with KF (0.6 mmol) and [18]crown-6 (0.6 mmol). The tube was evacuated at room

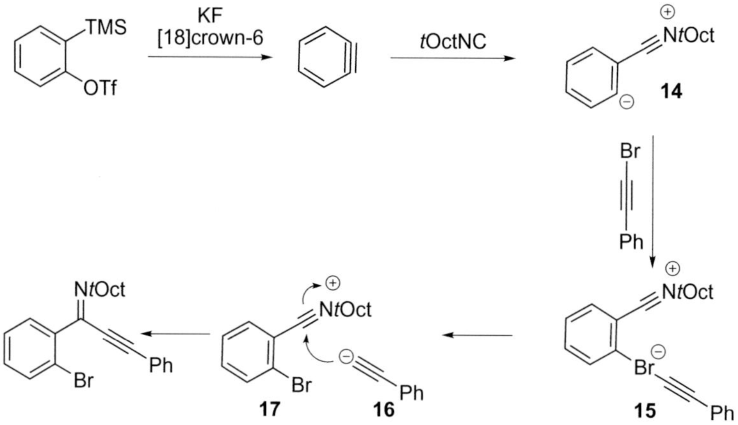

Scheme 3.17 Proposed mechanism

temperature for 1 h with stirring before addition of DME (1 mL) and a haloalkyne (0.15 mmol) under an argon atmosphere. Then an isocyanide (0.23 mol), and aryne precursor (0.3 mmol), and DME (1 mL) were added at 0 °C, and the resulting mixture was stirring at 0 °C. Upon completion, the reaction mixture was diluted with ethyl acetate and filtered through a Celite plug. The organic solution was washed with brine three times and dried over MgSO$_4$. Evaporation of the solvent and followed by recycling preparative HPLC gave the desired product.

As to the mechanism, the authors believed that this reaction could be triggered by the generation of zwitterion **14** (1,4-dipoles) from aryne and isocyanide. Subsequently, the zwitterion **14** underwent nucleophilic attack on the carbon-bromo bond of the bromoalkyne motif and provided the phenylacetylide **16** through the bromine ate complex **15**, followed by the carbon-carbon bond formation of **16** and the nitrilium cation **17** to furnish the final yn-1-imine product (Scheme 3.17).

3.1.1.3 Construction of C(sp)–C(sp^3) Bond

As the importance and diverse applications of alkyne units contained complex molecules, the development of efficient and sustainable methods for the construction of C(sp)–C(sp^3) bonds continues to be a challenging research topic in modern organic chemistry. In this context, the development of practical methods to construct C(sp)–C(sp^3) bonds with haloalkynes attracted considerable attention [73–79]. In this part, representative examples will be detailed discussed.

The coupling reaction of zinc-copper reagents with haloalkynes was one of most efficient strategy to construct C(sp)–C(sp^3) bonds. In 1998, Knochel's group reported that alkylborane could proceed transmetalation to the corresponding organozinc coumpound under the treatment of diisopropylzinc reagent, and subsequent transformation to the zinc-copper reagent with the addition of CuCN·2LiCl,

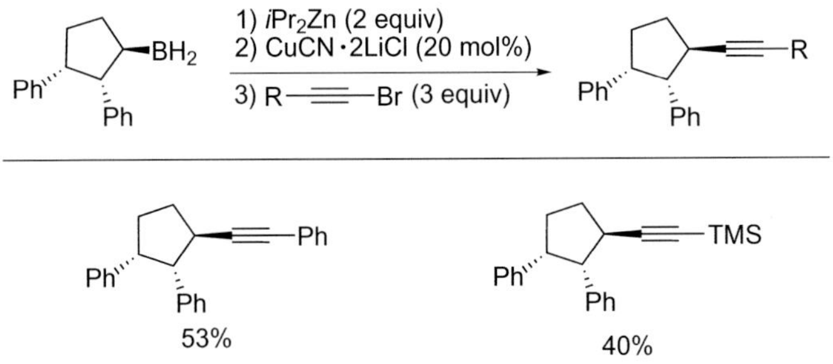

Scheme 3.18 Haloalkynes captured the zinc-copper reagent

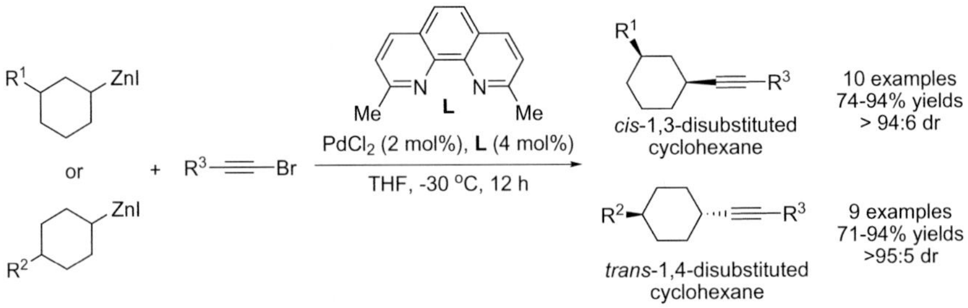

Scheme 3.19 Pd-catalyzed cross-coupling of cyclohexylzinc reagents with bromoalkynes

which could be captured by haloalkynes (Scheme 3.18) [80]. With the same strategy, Knochel [81], Williams [82], and Burton [83] respectively documented different zinc-copper reagents coupled with haloalkynes to construct C(sp)–C(sp^3) bonds.

General Procedure for Haloalkynes Captured the Zinc-Copper Reagent: A BH$_3$·THF solution (3 mmol) was slowly added to 1,2-diphenylcyclopentene (2 mmol) in THF (10 mL) at 20 °C. After 10 min, the resulting solution was heated at 50 °C for 3 h. The solvent and excess borane were removed under vaccum, and the residue was treated with a solution of *i*Pr$_2$Zn (4 mmol) in ether at 25 °C for 4 h. After removal of the solvent and excess of *i*Pr$_2$Zn under vacuum, the residue was diluted with THF (10 mL). The black precipitate of zinc was removed by filtration, and the filtrate was slowly treated at −90 °C with a solution of CuCN·2LiCl (20 mol%) in THF and after 15 min with haloalkyne (6 mmol) in THF. The reaction was allowed to warm to 25 °C and was quenched after 1 h with aq HCl (10 mL, 3 M) and extracted with ether. The crude product obtained after evaporation of the solvent was purified by chromatography.

Until 2011, the first Pd-catalyzed diastereoselective cross-coupling reaction of cyclohexylzinc reagents with bromoalkynes was reported by Knochel and co-workers (Scheme 3.19) [84]. Interestingly, the 3-substituted cyclohexylzinc reagent preferred to form *cis*-1,3-disubstituted cylcohexane derivatives, while 4-substituted cyclohexylzinc reagent favored to give *trans*-1,4-disubstituted cylcohexane derivatives. The high diastereoselectivity was assumed to be effected by a selective transmetalation step between the respective alkynyl(bromo)palladium complex and the cyclohexylzinc reagents, which led to the formation of the most thermodynamically stable palladium intermediates. Subsequently, reductive elimination proceeded with the retention of configuration and delivered the corresponding 1,3- and 1,4-disubstituted products. Later, they [85] and Baudoin [86] reported palladium-catalyzed cross-coupling reactions of haloalkynes with adamantylzinc reagents and α-zincated acyclic amines, respectively.

General Procedure for Pd-Catalyzed Cross-Coupling of Cyclohexylzinc Reagents with Bromoalkynes: A dry and N_2-flushed Schlenk tube (10 mL), equipped with a magnetic stirring bar and a septum, was charged with a solution of the respective alkynyl bromide (0.4 mmol), $PdCl_2$ (2 mol%) and neocuproine (4 mol%) in THF (1.5 mL) and cooled to −30 °C. A solution of the respective cyclohexylzinc iodide in THF (0.5 mmol) was slowly added at this temperature. The reaction mixture was stirred for 12 h. Then saturated aq. NH_4Cl solution (5 mL) was added. Phases were separated and the aqueous phase was extracted with Et_2O (20 mL × 3). The combined organic layers were washed with brine (10 mL) and dried over Na_2SO_4. The solvents were evaporated and the alkynylated product was purified via column chromatography on silica gel.

Grignard reagents could also coupled with haloalkynes [87]. In 2010, Cahiez reported the first efficient and practical copper-catalyzed alkynylation reaction of aryl and alkyl Grignard reagents [88]. This reaction has broad substrate scope and tolerates diverse functional groups, even tertiary alkyl Grignard reagent could also be used successfully. Notably, this reaction was highly chemoselective, and the key to obtain satisfactory yields was the slow addition of the Grignard reagents to the reaction mixture. Importantly, no Br/Mg exchange was observed in the arylation of alkynyl bromides (Scheme 3.20).

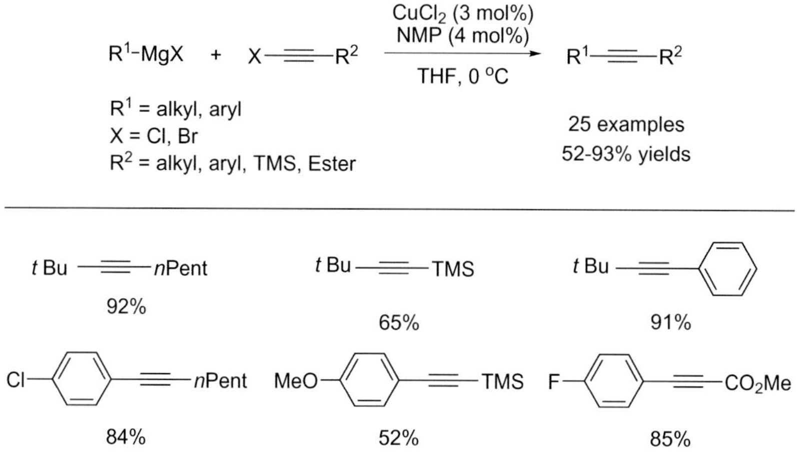

$$R^1\text{-MgX} + X\text{---}\!\!\equiv\!\!\text{---}R^2 \xrightarrow[\text{THF, 0 °C}]{\substack{\text{CuCl}_2\ (3\ \text{mol\%}) \\ \text{NMP}\ (4\ \text{mol\%})}} R^1\text{---}\!\!\equiv\!\!\text{---}R^2$$

R^1 = alkyl, aryl
X = Cl, Br
R^2 = alkyl, aryl, TMS, Ester

25 examples
52-93% yields

t Bu —≡— *n*Pent
92%

t Bu —≡— TMS
65%

t Bu —≡—⟨⟩
91%

Cl—⟨⟩—≡—*n*Pent
84%

MeO—⟨⟩—≡— TMS
52%

F—⟨⟩—≡—CO₂Me
85%

Scheme 3.20 Copper-catalyzed cross-coupling of Grignard reagents with haloalkynes

General Procedure for Copper-Catalyzed Cross-Coupling of Grignard Reagents with Haloalkynes: A dry and nitrogen flushed four-necked flask (100 mL) equipped with a mechanical stirrer, a thermometer, a nitrogen inlet, and a septum was charged with CuCl$_2$ (3 mol%), *N*-methylpyrrolidinone (4 mol%), haloalkyne (10 mmol), and THF (9 mL). After complete dissolution of the cuprous chloride (less than 30 min), the reaction mixture was cooled to 0 °C and a solution of Grignard reagent (12 mmol) was added with a syringe pump over a period of 45 min. At the end of the addition, stirring was continued for 30 min at 0 °C then the reaction was quenched with 1 N aqueous HCl solution (20 mL). The aqueous phase was extracted with diethyl ether (20 mL × 3). The combined organic layers were dried with MgSO$_4$, filtered, and concentrated under reduced pressure. The crude residue was purified by flash chromatography on silica gel.

On the basis of the experimental results and related literatures [89, 90], the authors proposed a reasonable mechanism in Scheme 3.21. The catalytic cycle was initiated by the generation of cuprate **18** from the Grignard reagent. Subsequently, the haloalkyne reacted with **18** to afford the vinylcopper reagent **21** through the complex **19/20** (carbocupration). Generally, **21** was generated via the reductive elimination of metallacyclopropene **19**. Finally, the unstable vinyl copper **21** underwent a β-halogen elimination to give the desired product and the organo-cooper **22**, which then reacted with another Grignard reagent to regenerate the cuprate **18**.

Besides organozinc reagents, Grignard reagents and zirconacyles [91], the more environmental friendly strategy to construct C(sp)–C(sp^3) bonds from haloalkynes was the utilization of activated C(sp^3)–H bonds. In 2007, Jørgensen and co-workers reported the first asymmetric direct alkynylation of cyclic β-ketoesters with haloalkynes under chiral phase transfer catalyst **23**. A large number of alkynylating reagents with chloride and bromide as the leaving groups and substituents such as alkyl and allyl esters, amides, ketones, and sulfones were demonstrated to be suitable substrates. Various cyclic β-ketoesters with different ring-sizes and also

Scheme 3.21 Proposed reaction mechanism

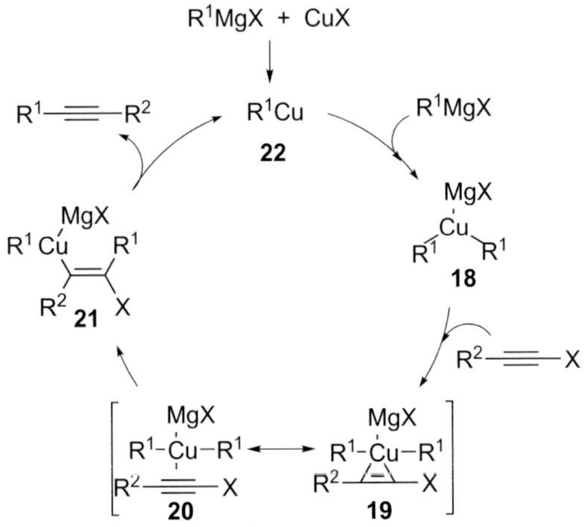

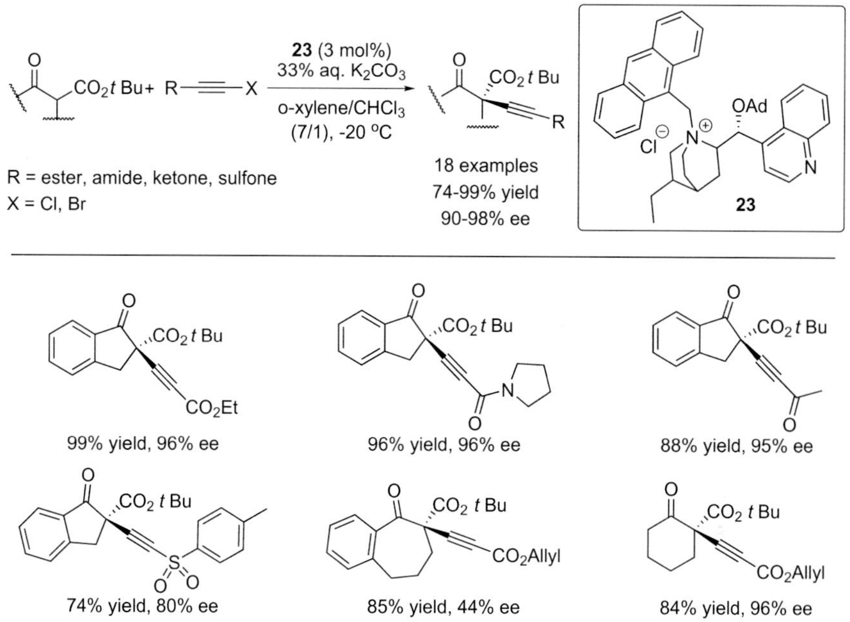

Scheme 3.22   Asymmetric alkynylation of cyclic β-ketoesters

including oxindoles were applicable to the standard reaction conditions. The corresponding optically active products were obtained in high yields with excellent enantioselectivities (Scheme 3.22) [92].

General Procedure for Asymmetric Alkynylation of Cyclic β-Ketoesters: To a sample vial equipped with a magnetic stirring bar was added β-ketoester (0.2 mmol), *o*-xylene/CHCl$_3$ (7:1, 1.3 mL), haloalkyne (0.26 mmol), and the catalyst **26** (3 mol%). The mixture was stirred for s short time at ambient temperature and was then placed at −20 °C. When the mixture had cooled, a cold solution of 33% aq. K$_2$CO$_3$ (0.6 mL) was added and the biphasic mixture was vigorously stirred. Upon completion, the organic phase was collected, and the aqueous layer was extracted with toluene two times. The combined organic fractions were loaded onto a chromatography column and the alkynylated product was obtained.

Importantly, the authors isolated and characterized the counterion of the catalyst and *p*-nitrophenolate by X-ray analysis, and they proposed a model of the catalyst-substrate intermediate which might explain the observed enantioselectivity of this organocatalytic enantioselective alkynylation reaction (Scheme 3.23). Alkali-metal enolate **24**, generated from the corresponding β-ketoesters by deprotonation with the bulk aqueous base, first underwent cation exchange with the chiral phase transfer catalyst **23**, which led to the organic soluable ammonium enolate **25** as a tight ion-pair. Due to the chiral environment provided by the ammonium motif, the enolate **25** added to the haloalkyne in a highly enantioselective manner, and formed the ammonium allenolate **26**. The allenolate **26** would undergo elimination

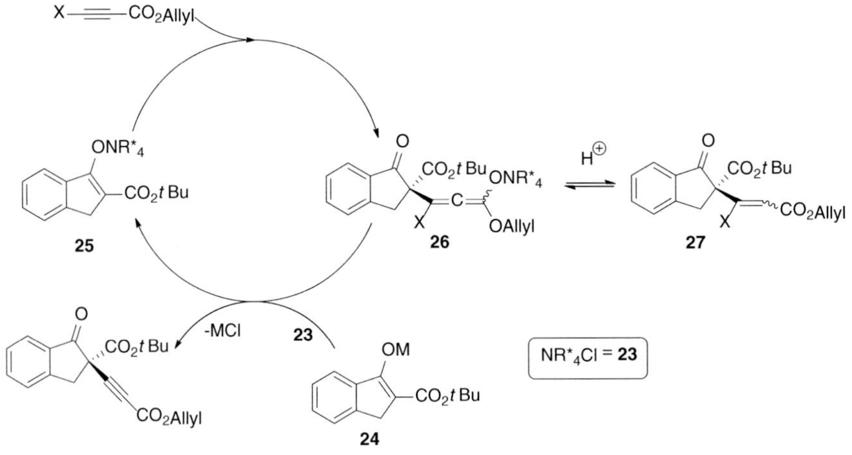
X—≡—CO2Allyl
ONR*4
CO2tBu
25
O
CO2tBu
ONR*4
C
X
OAllyl
26
H⊕
O
CO2tBu
X
CO2Allyl
27
O
CO2tBu
CO2Allyl
-MCl
23
OM
CO2tBu
24
NR*4Cl = 23

Scheme 3.23 The possible reaction mechanism

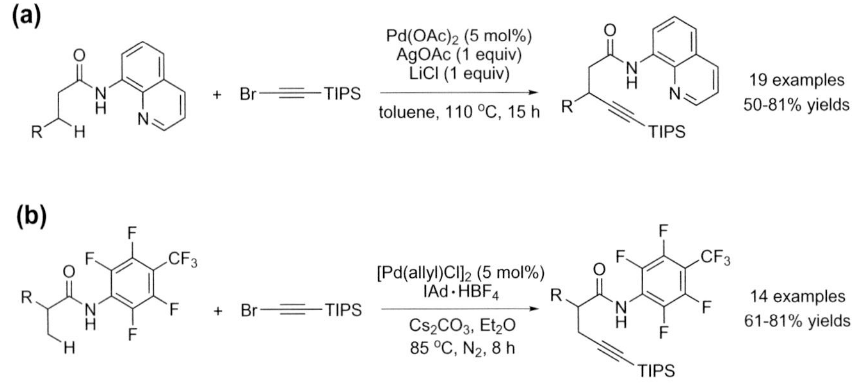
(a)
O
N
H
N
R
H
+ Br—≡—TIPS
Pd(OAc)2 (5 mol%)
AgOAc (1 equiv)
LiCl (1 equiv)
toluene, 110 °C, 15 h
TIPS
19 examples
50-81% yields
(b)
F
CF3
F
F
F
+ Br—≡—TIPS
[Pd(allyl)Cl]2 (5 mol%)
IAd·HBF4
Cs2CO3, Et2O
85 °C, N2, 8 h
TIPS
14 examples
61-81% yields

Scheme 3.24 Pd-catalyzed alkynlation of unactivated C(sp^3)–H bonds. **a** Chatani's work: Pd(II)/Pd(IV). **b** Yu's work: Pd(0)/Pd(II)

of X directly to deliver the desired alkynylated product or got protonated to afford trisubstituted vinylic ester **27**.

In 2011, taking advantages of C(sp^3)–H activation [93–95], Chatani [96] and co-workers documented the first alkynlation of unactivated C(sp^3)–H bonds via palladium(II/IV) process (Scheme 3.24a). Broad functional groups could be tolerated under the standard reaction conditions. Experiment results indicated that both the quinolone and the NH group were essential for the reaction. Two years later, Yu's group [97] reported a palladium(0)-catalyzed alkynlation of C(sp^3)–H bonds using Pd^0/NHC and Pd^0/PR_3 catalysts without the use of co-oxidants (Scheme 3.24b).

General Procedure for Pd-Catalyzed Alkynlation of Unactivated C(sp³)–H Bonds (Chatani's Work): To an oven-dried screw-capped vial (5 mL), *N*-(8-quinolinyl)hexanamide (0.5 mmol), (bromoethynyl)triisopropylsilane (0.75 mmol), Pd(OAc)$_2$ (5 mol%), AgOAc (0.5 mmol), LiCl (0.5 mmol) and toluene (1 mL) were added under a gentle stream of nitrogen. The mixture was stirred for 15 h at 110 °C and followed by cooling. The mixture was filtered through a Celite pad and concentrated in vacuo. The residue was subjected to column chromatography on silica gel (eluent: hexanes/Et$_2$O = 5/1 to 3/1) to afford the desired alkynylated product.

General Procedure for Pd-Catalyzed Alkynlation of Unactivated C(sp³)–H Bonds (Yu's Work): Substrate (0.1 mmol), [Pd(allyl)Cl]$_2$ (5 mol%), bis(adamantly) imidazolium tetrafluoroborate (0.02 mmol), and Cs$_2$CO$_3$ (0.2 mmol) were weighed in air and placed in Schlenk tube (50 mL) with a magnetic stir bar. The alkynyl bromide (0.2 mmol) and Et$_2$O (0.5 mL) were added, and the reaction vessel was evacuated and backfilled with nitrogen three times. The reaction mixture was first stirred at room temperature for 5 min and then heated to 85 °C for 8 h under vigorous stirring. Upon completion, the reaction mixture was cooled to room temperature. The solvents were removed under reduced pressure and the resulting mixture was purified by silica gel packed flash chromatography column using hexanes/EtOAc mixtures as the eluent.

Alternatively, the strategy of carbon-carbon bond cleavage could also be used to construct C(sp)–C(sp³) bonds [98]. In 2013, Martin and co-workers reported the reaction of Pd-catalyzed C(sp³)–C(sp³) bond cleavage of *tert*-cyclobutanols reacted with bromoacetylenes, which gave γ-alkynylated ketones in good yields (Scheme 3.25a) [99]. Later, Xu's group demonstrated the decarboxylative alkynylation of quaternary α-cyano acetate salts under copper catalysis (Scheme 3.25b) [100].

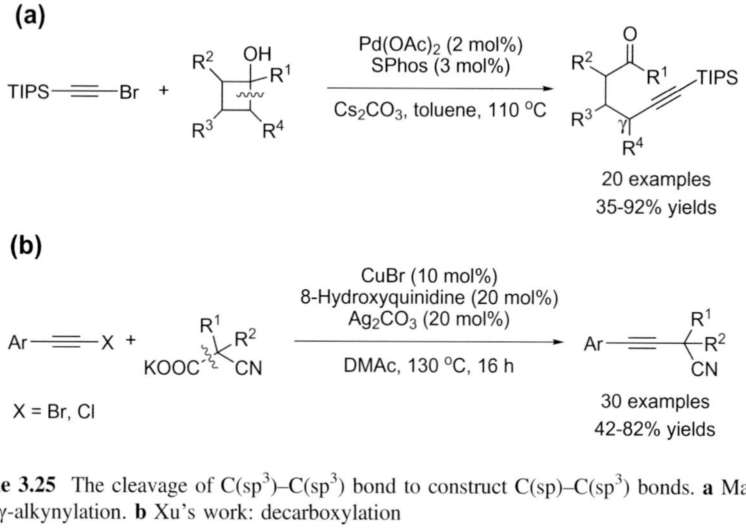

Scheme 3.25 The cleavage of C(sp³)–C(sp³) bond to construct C(sp)–C(sp³) bonds. **a** Martin's work: γ-alkynylation. **b** Xu's work: decarboxylation

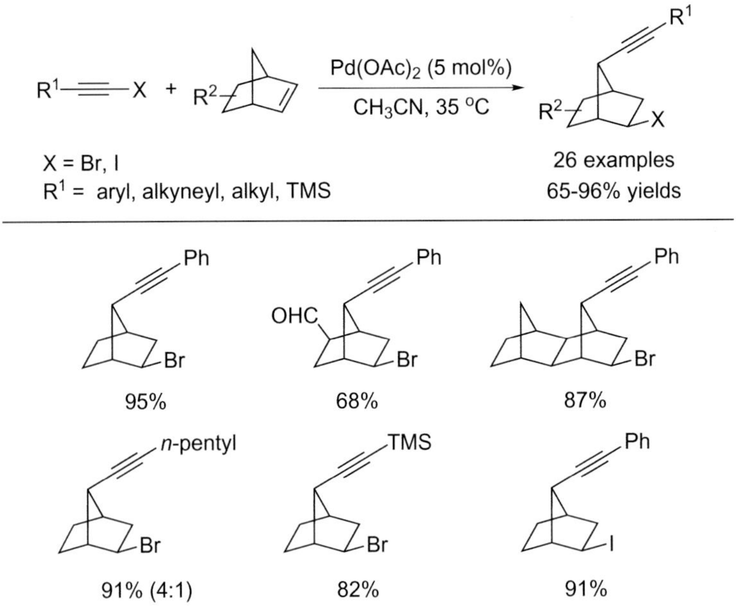

Scheme 3.26 Pd-catalyzed synthesis of 7-alkynyl norbornanes

However, all the previous reported method to construct $C(sp)$–$C(sp^3)$ bonds with haloalkynes, the halide motif was removed to the waste salts. In 2011, Jiang and co-workers demonstrated the first example of highly selective Pd-catalyzed intermolecular alkynylation reaction of norbornene derivatives, delivering diverse 7-alkynyl norbornane adducts that could not be easily accessed via traditional methods (Scheme 3.26) [101]. Outwardly, this unique transformation proceeded through the direct cleavage of the alkynyl–halogen bond, and followed by the constructions of $C(sp)$–$C(sp^3)$ and $C(sp^3)$–halogen bonds, featuring excellent atom economy. Their achievements in the synthesis of C7-functionalized norbornyl alkynes products proved the compatibility of nonclassical norbornonium cation with this catalytic system.

General Procedure for Pd-Catalyzed Synthesis of 7-Alkynyl Norbornanes: To a Schlenk tube (25 mL) was successively added Pd(OAc)₂ (5 mol%), CH₃CN (2 mL), norbornene (1.3 mmol) and haloalkyne (1 mmol). The resulting mixture was stirred at 30 °C for 10 h. Then, the mixture was filtered through a small amount of silica gel. The filtrate was concentrated under reduced pressure and the residue was purified by silica gel preparative TLC (*n*-hexane) to give the desired product.

Base on their experimental results and previous reports [33, 102, 103], they tentatively proposed the reaction mechanism (Scheme 3.27). Initially, the oxidative addition of Pd^0 or Pd^{II} species to haloalkyne generated a high-valent alkynylpalladium complex, followed by *cis*-insertion to give intermediate **28**. Subsequently, the bridging Pd complex **29** was formed, and then the Pd catalyst was transferred to the bridged carbon on the same side as the incoming alkyne, which led to the highly

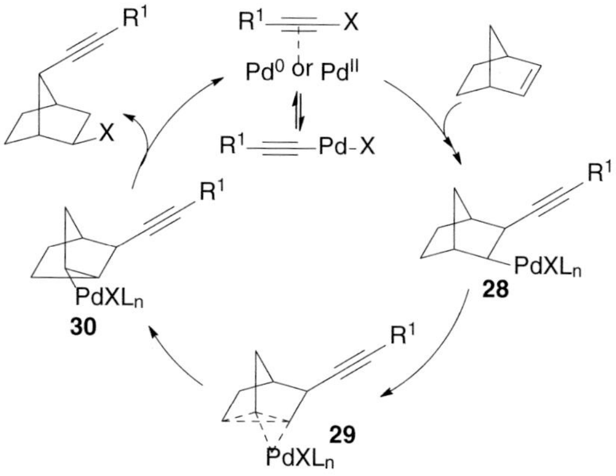

Scheme 3.27 Plausible
reaction mechanism

stereoselective formation of the alkylpalladium halide intermediate **30**. Finally, the
reductive elimination afforded the brominated product and regenerated the active
catalyst species.

Interestingly, Tong's group [104] also reported a Pd-catalyzed iodoalkynation of
norbornene with the employment of alkynyl iodides, which was found to be
strongly solvent dependent (Scheme 3.28). Polar solvents favored the unexpected
1,7-iodoalkynation adducts, while nonpolar solvents tended to the formation of
1,2-iodoalkynation products. The authors proposed a Pd^0/Pd^{II} reaction mechanism,
in which the formation of the product was relied on the solvent effects.

3.1.2   Construction of Carbon-Nitrogen Bond

Ynamines and ynamides are modern functional motifs with increasing significance
that can easily and efficiently transfer to the nitrogen-containing compounds, providing access to privileged scaffolds widely existed in natural products, bioactive
molecules and functional materials [105–107]. Taking advantages of the development of efficient methods for ynamide's preparation, the chemistry of ynamide has
experienced rapid expansion during the past decade [108–111]. Particularly, the
amidative cross-coupling of haloalkynes and amines has emerged as one of the
most important strategies. Although the first example of ynamides were reported by

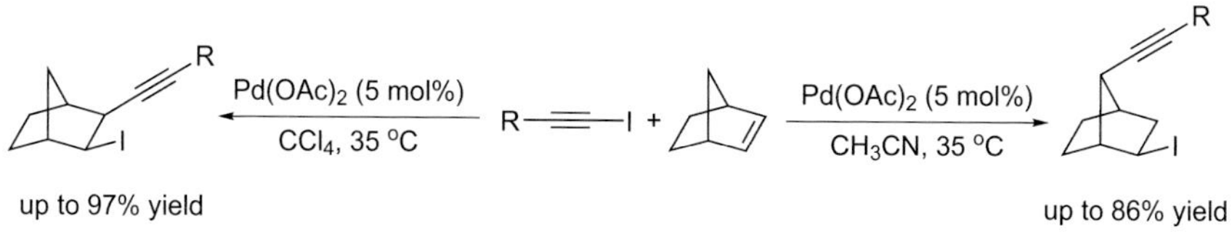

Scheme 3.28 Pd-catalyzed iodoalkynation of norbornenes

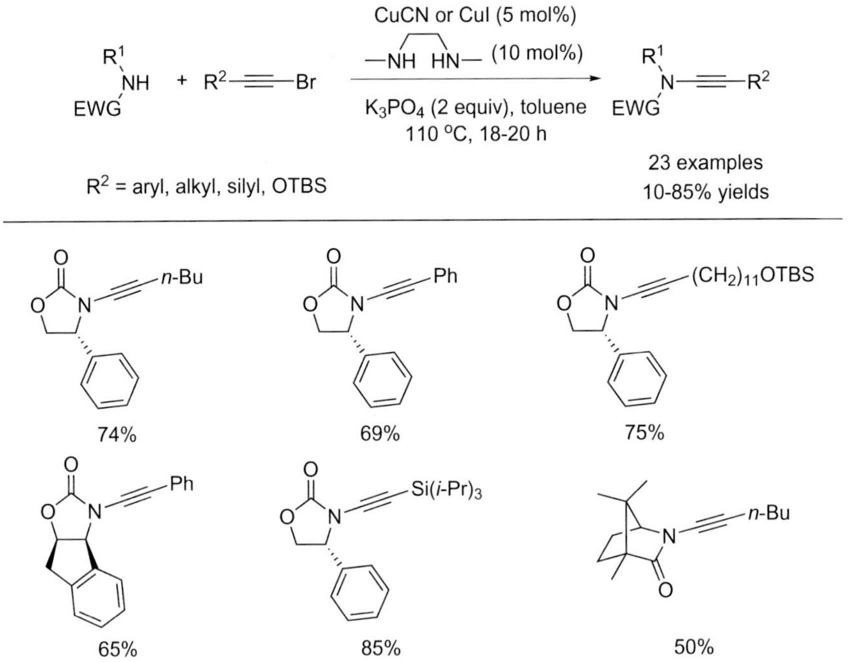

Scheme 3.29 Cu-catalyzed ynamide formation reactions

Viehe in 1972 [112], and the first synthesis of ynamides through metal-mediated reactions was documented in 1985 by Balsamo and Domiano [113]. However, limited progress [114] was achieved until 2003, Hsung et al. [115] disclosed the first Cu-catalyzed ynamide formation reaction, which provided a straightforward and atom-economical access to various ynamides (Scheme 3.29). Generally, CuCN led to more consistent results overall, although no significant different results appeared when CuI was used instead of CuCN. This coupling reaction tolerated various types of haloalkynes, and provided a direct entry to chiral ynamides in good yields. Later on, they developed a more efficient and practical catalytic system for ynamides synthesis, with the utilization of inexpensive $CuSO_4 \cdot 5H_2O$ as catalyst and 1,10-phenanthroline as the ligand [116, 117]. This protocol had a broad functional group tolerance and was also applicable for intramolecular amidation reactions, which could be applied to the construction of unique macrocyclic yna-mides that contained up to 19-membered ring system (Scheme 3.30). Except for Cu salts [118–125], other transition metals also presented their high reactivity for the synthesis of ynamides. In 2009, Zhang's group disclosed the first Fe-catalyzed coupling of amides and alkynyl bromides [126]. It was announced that $FeCl_3 \cdot 6H_2O$, an environmentally friendly alternative to Cu salt, was also a practical and efficient catalyst for ynamide synthesis (Scheme 3.31).

General Procedure for the Cu-Catalyzed Ynamide Formation Reactions: To a reaction vial was added amide (1 mmol), K_3PO_4 (2 mmol), and CuCN (5 mol%).

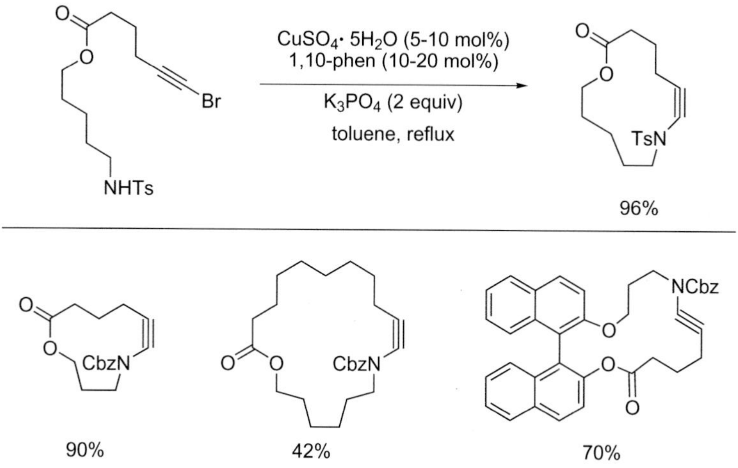

Scheme 3.30 Construction of unique macrocyclic ynamides

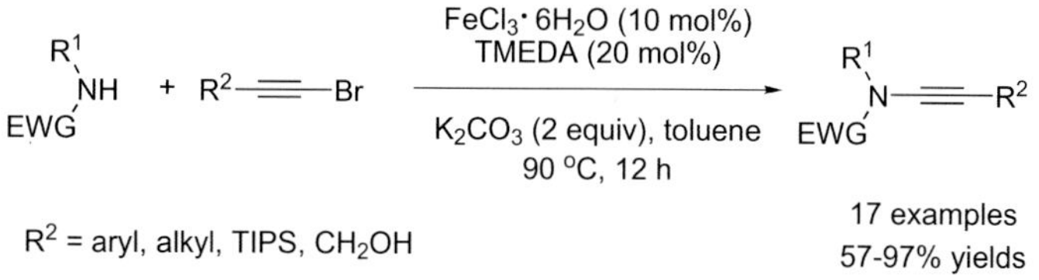

Scheme 3.31 Fe-catalyzed coupling of amides and bromoalkynes

Bromoalkyne (1 mmol) was then added in a solution of anhydrous toluene (10 mL) followed by addition of *N,N*′-dimethylethylene diamine (0.1 mmol). The reaction vial was sealed and placed in an oil bath at 110 °C for 15–24 h. The reaction was followed with TLC, LCMS, and/or GCMS analysis. Upon completion, the reaction mixture was filtered through a small bed of silica gel and concentrated under vacuum. Purification of the residue by silica gel chromatography (gradient eluent: 0–50% EtOAc in hexane) afforded the corresponding ynamide products.

3.1.3 Construction of Carbon-Sulfur Bond

Acetylenic thioethers are an important class of compounds [127–129]. However, multi-step synthesis is usually involved for their preparation [130]. In 1962, Miller and co-workers developed a simple process to synthesize acetylenic thioethers from haloalkynes and sodium thiolates via nucleophilic subsitution at an acetylenic carbon (Scheme 3.32) [131]. The key factor for the success of this operation was the utilization of aprotic solvent DMF as the solvent. Interestingly, the nucleophilic displacement on these haloalkynes proved to be surprisingly facile even at −25 °C.

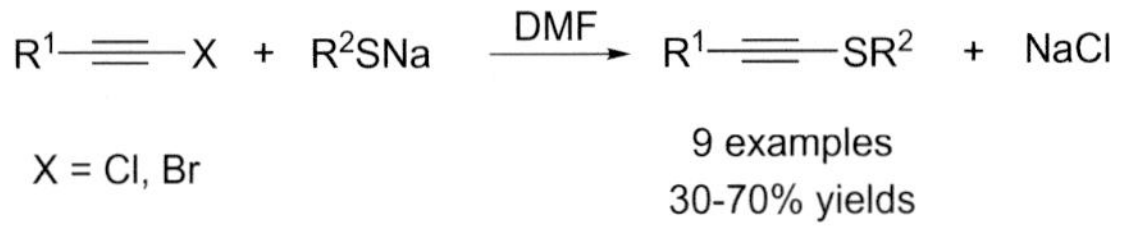

$$R^1\!\!-\!\!\equiv\!\!-X \ + \ R^2SNa \ \xrightarrow{\ DMF\ } \ R^1\!\!-\!\!\equiv\!\!-SR^2 \ + \ NaCl$$

X = Cl, Br

9 examples
30-70% yields

Scheme 3.32 Synthesis of acetylenic thioethers

Importantly, it proved that the nucleophilic substitution at an acetlyenic carbon is possible.

General Procedure for the Synthesis of Acetylenic Thioethers: The solution of sodium thiolate (1.02 equiv) and haloalkyne (1.0 equiv) in DMF (0.12 mmol/mL) was mixed and stored at −30 °C in a stoppered flask which had been flushed with nitrogen. If the reaction was slow, the temperature of the solution was raised to 25 °C or higher if need be. Unnecessary heating appeared to reduce the yields of products. On completion of the reaction, the solution was treated with ice and water and extracted with ether to give the impure sulfides. Careful distillation gave the acetylenic thioether products.

3.1.4   Construction of Carbon-Phosphorus Bond

Alkynyl-phosphorus compounds are an important class of triple bond-containing, extremely versatile chemicals in modern synthetic chemistry, which are broadly available for the preparation of structurally sophisticated phosphorus-containing compounds [132]. In this context, the preparation of alkynyl-phophorus compounds has attracted considerable attention over the past decades [133]. In 2014, Gao and co-workers developed Cs_2CO_3-promoted one-pot synthesis of alkynylphophorus from bromoalkynes or 1,1-dibromo-1-alkenes via carbon-phosphorus bond formation. Without base, 1,1-dibromo-1-alkenes could not convert to the desired product under the standard reaction conditions [Scheme 3.33, Eq. (1)]. Mechanism investigation indicated that 1,1-dibromo-1-alkene could be transferred to the corresponding bromoalkyne under the treatment of base, and bromoalkyne was the reactive species [Scheme 3.33, Eqs. (2) and (3)]. Subsequently, the addition of triethyl phosphate led to the formation of quaternary phosphonium salt **31**, with the release of bromide group affording the phosphonium salt **32**, which then underwent Michaelis-Arbuzov type reaction to afford the alkynyl-phosphorus products (Scheme 3.33) [134].

General Procedure for the Synthesis of Alkynyl-Phosphorus Compounds: An oven-dried Schlenk tube with Cs_2CO_3 (0.75 mmol) was evacuated and purged with argon three times. A mixture of 1,1,-dibromo-1-alkene or bromoalkyne (0.5 mmol) and *P*-nucleophiles (0.55 mmol) in toluene (1.5 mL) was added to the tube and stirred at 120 °C for 24 h. The suspension was filtered and washed with EtOAc (5 mL × 3). The combined solvent was removed under reduce pressure. The residue was purified by silica gel chromatography using a mixture of petroleum ether and ethyl acetate as eluent.

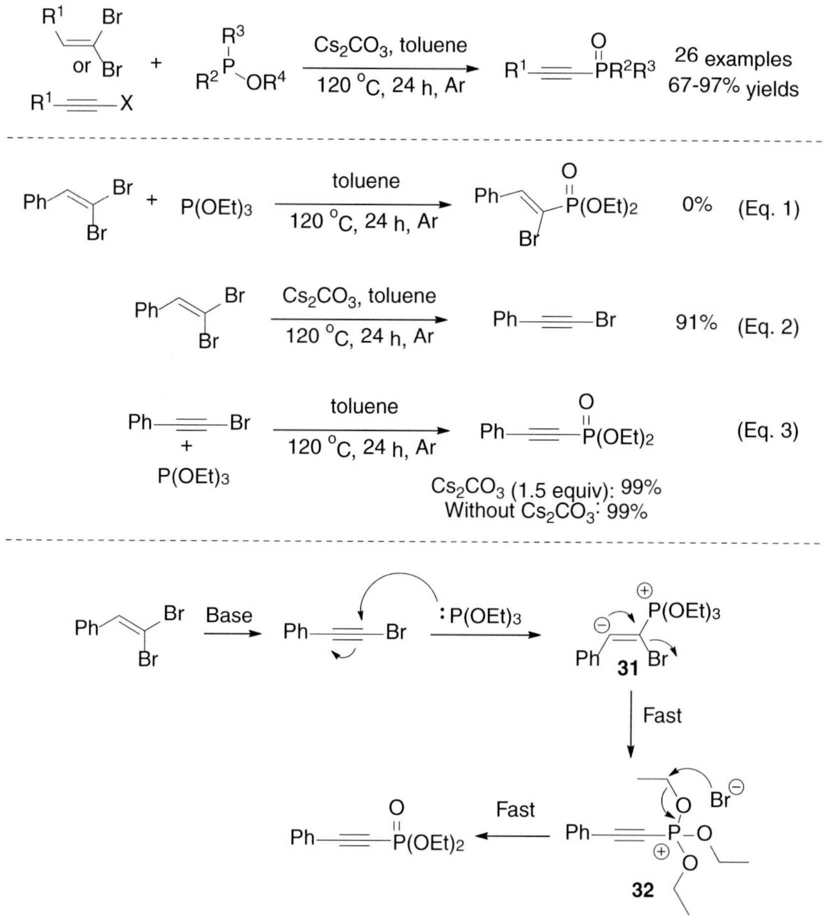

Scheme 3.33 Synthesis of alkynyl-phosphorus compounds

3.2 Transformations of Carbon-Carbon Triple Bond Motif

Alkyne motif is one of the most reactive and useful fundamental units in synthetic chemistry, which exhibits rich and tunable reactivity particularly under the treatment of transition metal catalysts. Consequently, the diverse transformations of alkyne motif contained molecules have attracted considerable attention during the past decades [135]. Accordingly, transformations of haloalkyne reagents based on the highly reactive carbon-carbon triple bond, could realize the facile synthesis of frameworks that previously were unable or difficult to obtain.

3.2.1 Nucleophilic Additions

Due to the central role of heteroatom-contained olefins in biological systems and pharmaceutical applications, the development of efficient and sustainable methods to synthesize this class of compounds is a long-term task in the area of synthetic and medicinal chemistry [136–139]. Among various protocols to achieve this goal, the nucleophilic addition of haloalkynes represents a series of reactions with important synthetic value to construct $C(sp^2)$–X bonds.

3.2.1.1 Halogen Nucleophiles

Dihaloalkenes have emerged as one of the most versatile intermediates in organic synthesis, especially in the transition metal-catalyzed cross-coupling reactions. However, the traditional methods for the preparation of dihaloalkenes usually suffer some limitations, such as poor selectivity and difficult purification [140–142]. In 2010, Jiang's group documented the first example of a facile two-step synthesis of (Z)-2-halo-1-iodoalkenes from simple terminal alkynes, delivering the desired products in moderate to excellent yields with high regio- and stereoselectivities (Scheme 3.34) [143]. This method was transition-metal free, and exhibited excellent functional group compatibility. Additionally, the useful halo-iodoalkene adducts could be easily transformed to the conjugated (Z)-haloenynes and

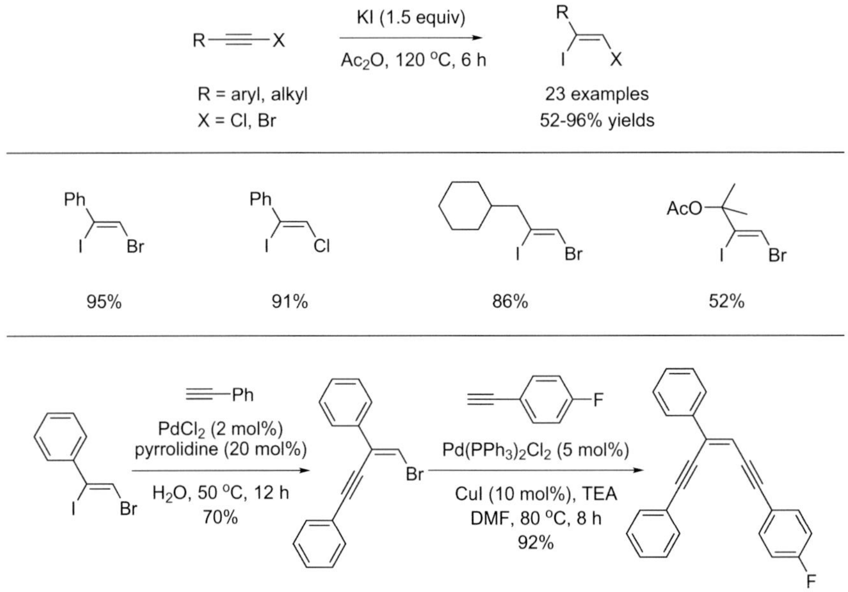

Scheme 3.34 Halogenation reaction of haloalkynes

asymmetrical (Z)-enediynes in good yields through selective Sonogashira coupling pathway. Later in 2012, Zhu and co-workers realized the hydrohalogenation of alkynyl halides to construct (Z)-1,2-dihaloalkenes under palladium catalysis [144].

General Procedure for the Halogenation Reaction of Haloalkynes: The mixture of haloalkyne (1 mmol), KI (1.5 mmol) and acetic anhydride (1.5 mL) were heated at 120 °C for 6 h. Then, the mixture was allowed to cool to room temperature, and water was added. The resulting mixture was extracted with ethyl acetate (15 mL × 3), and the combined extract was dried with anhydrous $MgSO_4$. The solvent was removed under reduced pressure and the residue was separated by column chromatography to give the desired dihaloalkenes.

Due to the unique physical and biological properties of fluorinated molecules, the corresponding vinyl fluoride products were quite attractive for synthetic and medicinal chemists [145–149]. Although the halide nucleoaddition to haloalkynes have provided a diverse set of haloalkene derivatives, the incorporation of fluorine atom into the final olefin products through transition metal catalysis is still a challenging target. Rare effective methods are available for the transition metal-catalyzed direct synthesis of simple fluoroalkene derivatives without additional functional sites [150]. In 2012, Jiang and coworkers revealed a one-pot silver-assisted regio- and stereoselective bromofluorination reaction of terminal alkynes (Scheme 3.35) [151]. The corresponding bromofluoroalkenes could be obtained in high yields with excellent selectivity. It was found that the electron-rich internal carbon-carbon triple bond was tolerated under the standard reaction conditions. To gain further insight into the mechanism of the catalytic cycle, the authors conducted some control experiments, such as the direct fluorination reactions of haloalkynes. Gratifyingly, both bromoalkynes and chloroalkynes exclusively afforded the fluorinated products in good yields. However, the iodoalkynes transformed to the corresponding iodofluoroalkene and diiodofluoroalkene adducts in a ratio of 2:1, due to the higher reactivity of iodoalkynes than bromoalkynes and chloroalkynes. Notably, the stereoselective functionalization of bromide subunit was successfully realized via Sonogashira or Suzuki coupling reactions. Thus, the present synthetic protocol would be applicable to obtain the 1-fluoro-1,3-enyne molecules that widely exist in numerous organic materials and biologically active compounds.

General Procedure for Ag-Assisted Bromofluorination Reaction of Terminal Alkynes: To a Schlenk tube was successively added NBS (1.1 mmol), AgF (2.5 mmol), CH_3CN (wet, 2 mL), and alkyne (1 mmol). The resulting mixture was stirred at 80 °C for 10 h. Then, the mixture was allowed to cool to room temperature, and filtered through a small amount of silica gel. The filtrate was concentrated under reduced pressure and the residue was purified by silica gel preparative TLC (*n*-hexane) to give the desired product.

According to the previous literatures [152, 153] and the obtained experimental results, the authors tentatively proposed the possible reaction mechanism (Scheme 3.36). Initially, the bromoalkyne intermediate was formed through the Ag-promoted bromination of terminal alkynes. Subsequently, the Ag cation was attacked by the triple bond of bromoalkyne to give a π-complex **33**, which was then

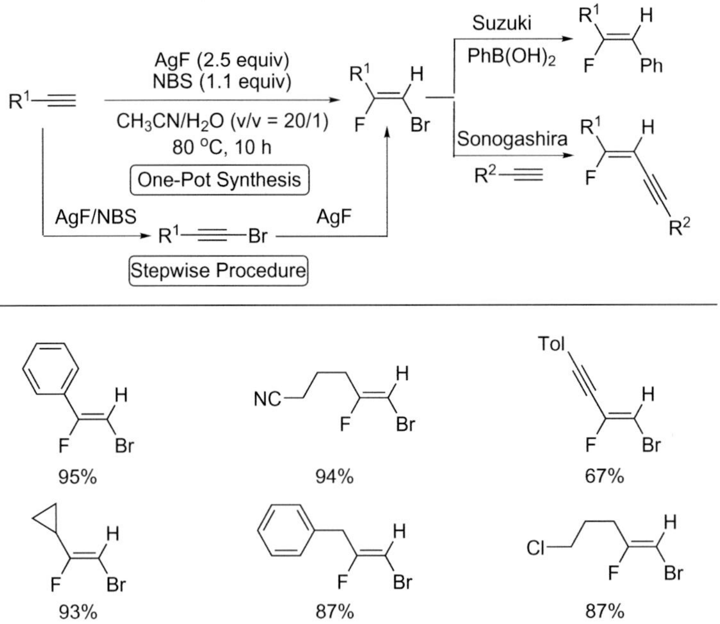

Scheme 3.35 Ag-assisted bromofluorination reaction of terminal alkynes

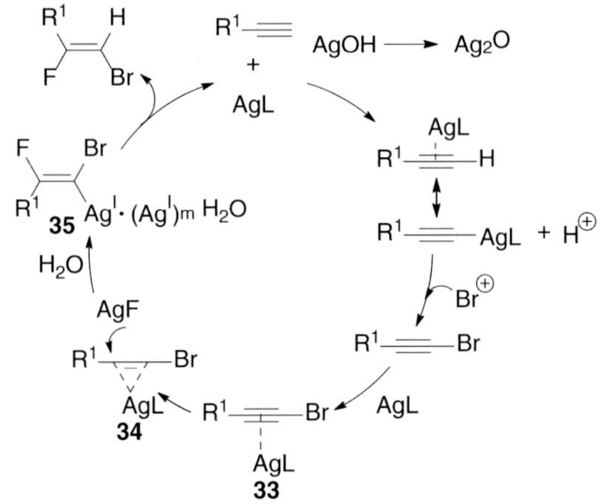

Scheme 3.36 Proposed mechanism

transferred to the corresponding vinylsilver intermediate **35** by *trans*-addition of AgF to bromoalkyne. Finally, protonation of **35** afforded the final product and silver oxide. The high regio- and stereoselectivities were proposed to be originated from the back-side attack of the fluoride anion (**34** to **35**) as well as the bromide atom was regarded as both an activating and regio-directing functional group. However, another mechanism involving the formation of vinylsilver intermediate **35** through the nucleophilic addition of fluoride to bromoalkyne could not be ruled out.

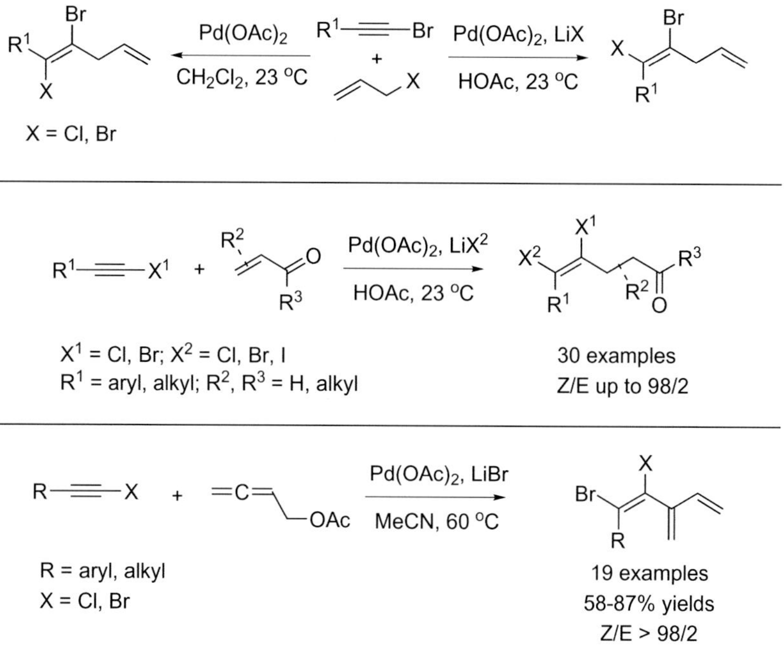

Scheme 3.37 Pd-catalyzed synthesis of haloalkenes

Halogenated 1,*n*-dienes, another significant kind of structural building blocks, are usually employed to construct biologically active and multifunctional compounds [154–157]. In 2011, Zhu's group documented an efficient and selective method for the synthesis of (1E)- or (1Z)-1,2-dihalo-1,4-dienes via Pd-catalyzed coupling of haloalkynes and allylic halides (Scheme 3.37) [158]. Interestingly, the E/Z selectivity of the diene product could be switched by the addition of stoichiometric lithium halides. With the same halopalladation strategy, they also reported a Pd-catalyzed coupling approach of alkynyl halides with α,β-unsaturated carbonyls [159] and 2,3-butadienyl acetates [160] for the synthesis of *cis*-1,2-dihaloalkene and (1Z)-1,2-dihalo-3-vinyl-1,3-diene derivatives.

In 2013, Jiang's research group reported Pd-catalyzed intermolecular cross-coupling reactions for the stereoselective synthesis of functionalized 1,*n*-dienes in ionic liquids (ILs) [161]. The ionic liquids not only acted as a solvent in the reaction, but also served as the excess halide ions source to control the Z/E selectivity. A chain-walking mechanism for this transformation is tentatively proposed in Scheme 3.38. Firstly, Pd complex was formed in situ in ILs and vinyl-palladium intermediate **36** was generated by *trans*-halopalladation of the alkyne moiety in the presence of excess halide ions in a polar solvent system. Subsequently, **36** underwent alkene insertion to deliver the alkylpalladium species **37**, followed by rapid β-H elimination and reinsertion to change the position of the metal on the alkyl chain, affording the intermediate **38**. Finally, a β-heteroatom elimination gave the obtained dihalo-1,*n*-diene adduct.

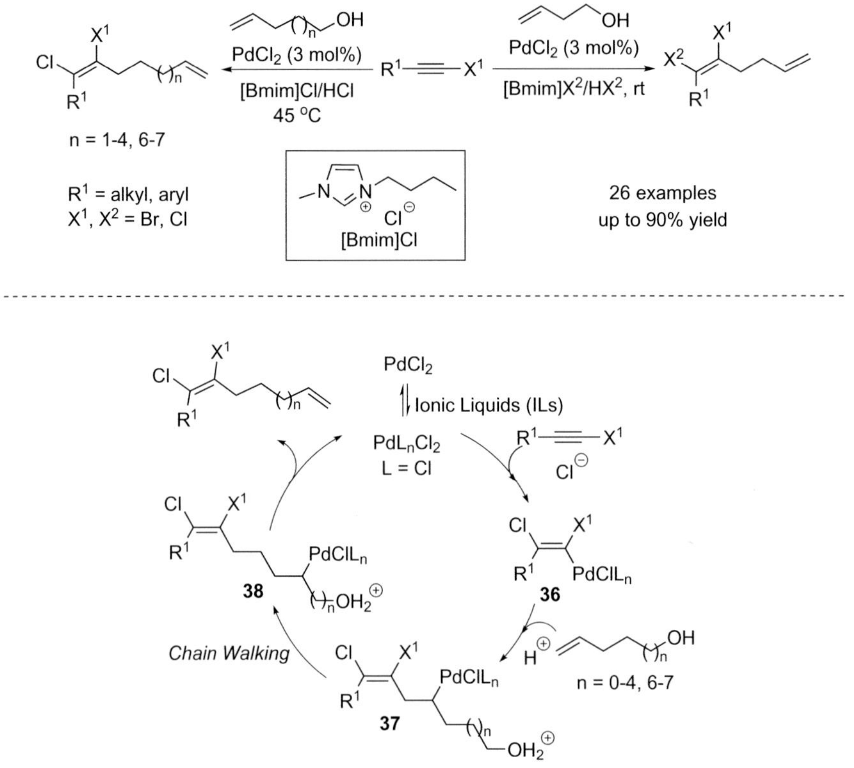

Scheme 3.38 Pd-catalyzed synthesis of functionalized dihalo-1,*n*-dienes

General Procedure for Pd-Catalyzed Synthesis of Functionalized Dihalo-1,n-dienes: To a test tube (10 mL) equipped with a magnetic stirring bar were successively added haloalkyne (0.5 mmol), alcohol (0.6 mmol), palladium chloride (3 mol %), ionic liquid (0.5 mL), HX (X = Cl, Br) (0.25 mL). The mixture was stirred under the atmosphere of air at room temperature. After the reaction was completed, the mixture was poured into ethyl acetate (30 mL). The organic layer was washed with brine to neutral, dried over anhydrous MgSO$_4$, concentrated in vacuum. Purification of the residue on a preparative TLC afforded the desired products.

Additionally, saturated lactones are found in a wide range of synthetically challenging and biologically significant natural products, which exhibit extraordinary pharmaceutical and biological properties [162–164]. Taking the advantages of halo-nucleopalladation, Jiang and coworkers realized the first example of palladium-catalyzed intermolecular cascade annulation for the construction of γ-lactones with regio- and stereoselectivity in ionic liquids (ILs) (Scheme 3.39) [165]. Besides the broad substrate scope, their cascade annulation reaction tolerated diverse functional groups. Significantly, all the obtained products were resulted from *trans* addition under the standard reaction conditions. Interestingly, they also applied

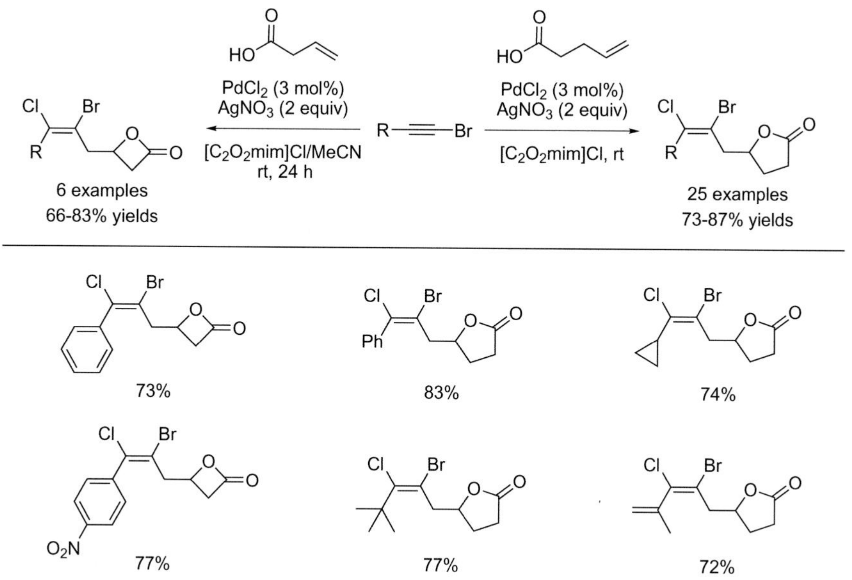

Scheme 3.39 Palladium-catalyzed synthesis of β- and γ-lactones

their method to the reaction with but-3-enoic acid, and various β-lactones could be successfully obtained under similar reaction conditions (Scheme 3.39) [165].

General Procedure for Palladium-Catalyzed Synthesis of β-, and γ-Lactones: To a test tube (10 mL) equipped with a magnetic stirring bar were successively added haloalkyne (0.25 mmol), the corresponding acid (0.3 mmol), palladium chloride (3 mol%), ionic liquid (0.5 mL). The mixture was stirred under the atmosphere of air at room temperature. After the reaction was completed, the mixture was poured into ethyl acetate (30 mL). The organic layer was washed with brine to neutral, dried over anhydrous MgSO$_4$, concentrated in vacuum. Purification of the residue on a preparative TLC afforded the lactone product.

Based on the current results and previous literatures [161, 166, 167], the authors proposed the possible reaction mechanism, which was illustrated in Scheme 3.40. Firstly, Pd complex was formed *in situ* in ILs and vinylpalladium intermediate **36** was generated by *trans*-halopalladation of the alkyne moiety in the presence of excess halide ions in a polar solvent system. Subsequently, **36** underwent alkene insertion. The vinylpalladium species coordinated to the oxygen atoms of the hydroxyl group to generate the palladium/alkyl intermediate **39**. Finally, a reductive elimination gave the obtained lactone adduct and Pd^0. Noteworthy, a silver mirror was observed after the completion of the reaction. Hence, the resulting Pd^0 was further oxidized to Pd^{II} which would be involved the next catalytic cycle.

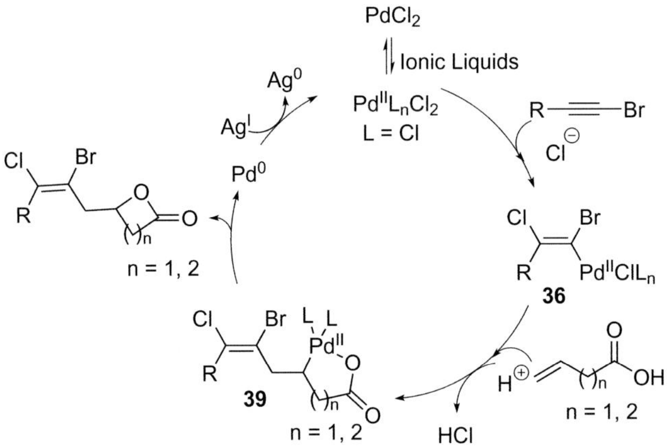

Scheme 3.40 Possible mechanism for palladium-catalyzed synthesis of lactones

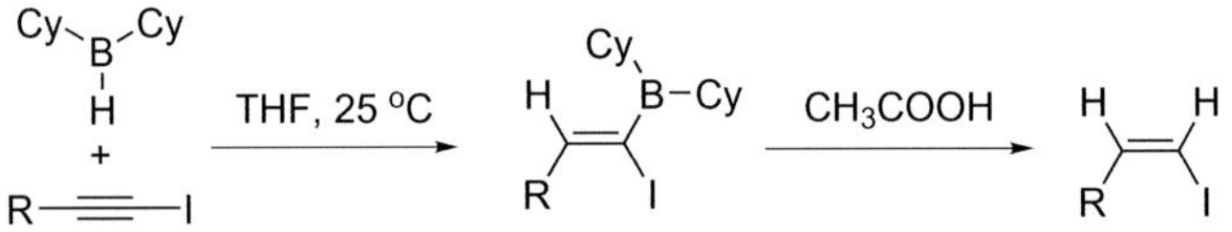

Scheme 3.41 Synthesis of *trans*-α-halovinylboranes

3.2.1.2 Boron Nucleophiles

As the diverse transformation abilities of organoborane compounds, the synthesis of α-halovinylboranes attracted the attention of scientists early in 1967 [168]. Addition of dicyclohexylborane to haloalkynes afforded the corresponding *trans*-α-halovinylboranes, which could directly convert to the corresponding ketones. Importantly, it was found out that *trans*-α-halovinylboranes were stable toward alkyl group migration in THF solvent as evidenced by their conversion into *cis*-vinyl halides upon hydrolysis with acetic acid (Scheme 3.41). With this protocol of α-halovinylboranes, Brown [169] and Walsh [170] applied them into diverse transformations.

General Procedure for the Synthesis of trans-α-Halovinylboranes: To a suspension of dicyclohexylborane (30 mmol) in THF (60 mL) at 0 °C was added haloalkyne (30 mmol). The reaction mixture was maintained for an additional 30 min at 20–30 °C, and then used directly for the next transformation.

3.2.1.3 Carbon Nucleophiles

The addition of carbon necleophiles to unsaturated bonds is a very important strategy to construct carbon-carbon bond. In 2008, Nakamura and co-worker

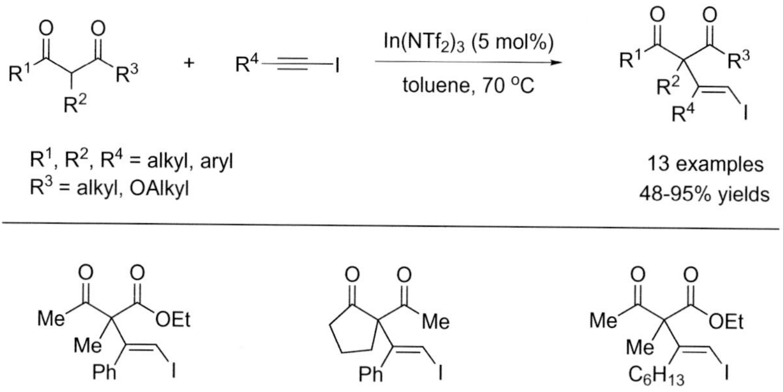

Scheme 3.42 Indium-catalyzed addition of 1,3-dicarbonyl compounds to 1-iodoalkynes

reported indium-catalyzed addition of 1,3-dicarbonyl compounds to 1-iodoalkynes [171]. This reaction proceeded exclusively in a *syn*-fashion to give E-alkenyl iodide in high yields. And the structure of the product was unambiguously determined by X-ray crystallographic analysis. Importantly, the iodine atom not only served as an activating group, but also as a direct group that controlled the regioselectivity of the addition (Scheme 3.42). Later, Jiang's group [172] documented the nucleophilic addition of isocyanides to bromoalkynes via palladium catalysis, and Vadola [173] realized the gold-catalyzed dearomative spirocyclization of aryl alkynoate esters.

   General Procedure Indium-Catalyzed Addition of 1,3-Dicarbonyl Compounds to 1-Iodoalkynes: A mixture of 1,3-dicarbonyl compound (2 mmol), 1-iodoalkyne (3 mmol), and In(NTf$_2$)$_3$ (5 mol%) in toluene (2 mL) was heated in the dark at 70 °C for 4 h. The mixture was filtered through a pad of silica gel and concentrated. The crude product as purified by silica gel column chromatography to give the desired product.

3.2.1.4 Nitrogen Nucleophiles

β-Halo enamines are not only important building blocks in functional molecules, but also reactive intermediates for many chemical processes [174–176]. Undoubtedly, the addition of nitrogen to haloalkynes is a convenient route to facile access β-halo enamines. In 2013, Wang [177] and co-workers reported the silver-catalyzed addition reaction of tetrazoles with bromoalkynes, delivering the β-halo enamine products in good yields and excellent stereoselectivities. Control experiments indicated that *N*-phenylcyanamide (**40**) was the reactive intermediate (Scheme 3.43). Importantly, the β-halo enamine products could be further transformed to 2-arylindoles.

   General Procedure for Ag-Catalyzed Synthesis of β-Halo Enamines: A reaction tube (10 mL) was charged with tetrazole (0.5 mmol), bromoalkyne (0.75 mmol), Ag$_2$O (20 mol%) and DMSO (2 mL). The reaction vessel was placed in an oil bath.

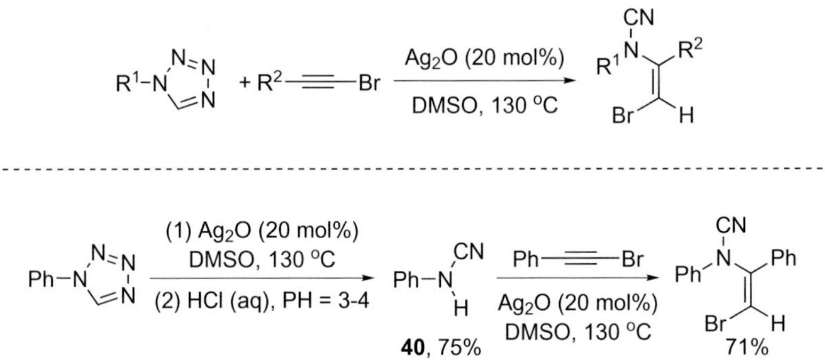

Scheme 3.43 Ag-catalyzed synthesis of β-halo enamines

After the reaction was carried out at 130 °C for 12 h, it was cooled to room temperature, extracted with EtOAc (5 mL × 3). The organic layers were combined, dried over MgSO₄, and concentrated. The residue was purified by flash chromatography on silica gel to give the β-halo enamine product.

3.2.1.5 Oxygen Nucleophiles

The β-haloenol acetate subunits are of considerable significance in organic synthesis and pharmaceutical chemistry [178–180]. It is striking, however, very few catalytic methods have been developed to construct the OC=CX motif in one step from simple terminal alkynes [181, 182]. In 2010, Jiang's group reported the first example of Ag-catalyzed alkyne difunctionalization reaction to afford the (Z)-β-haloenol acetate derivatives with extremely high regio- and stereoselectivities (Scheme 3.44) [183]. They proposed that the haloalkyne intermediate was first generated and then the triple bond attacked the Ag cation to deliver a π-complex **41**, which was transferred to the corresponding σ-complex **42** through the nucleophilic attack of acetic anion. Finally, protonation of **42** gave the desired β-haloenol acetate product. Accordingly, the high regio- and stereoselectivities might be owing to the stabilization effect of halogen atom to the Ag catalyst. Later, plenty of methods have been developed for the nucleophilic addition of haloalkynes with diverse oxygen nucleophiles, and delivered the corresponding β-haloenol [184, 185] or α-haloketone [186, 187] derivatives in good yields.

General Procedure for Ag-Catalyzed Synthesis of (Z)-β-Haloenol Acetates: To a test tube (10 mL) equipped with a magnetic stirring bar were successively added terminal alkyne (1 mmol), NBS (1.2 mmol), acetic anhydride (2 mL), silver tetrafluoroborate (5 mol%). The mixture was stirred at 120 °C for 12 h. Then the solution was allowed to cool to room temperature, extracted with ethyl acetate (15 mL × 3). The combined extracts were dried over anhydrous MgSO₄, filtered and concentrated in vacuum. The residue was purified by column chromatography to give the haloenol acetate product.

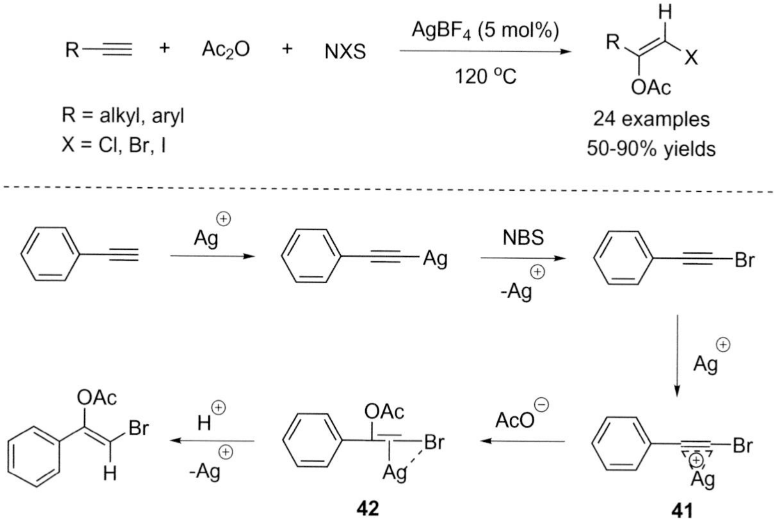

R——≡ + Ac₂O + NXS

$$\xrightarrow[\text{120 °C}]{\text{AgBF}_4\ (5\ \text{mol\%})}$$

R = alkyl, aryl
X = Cl, Br, I

24 examples
50-90% yields

Scheme 3.44 Ag-catalyzed synthesis of (Z)-β-haloenol acetates

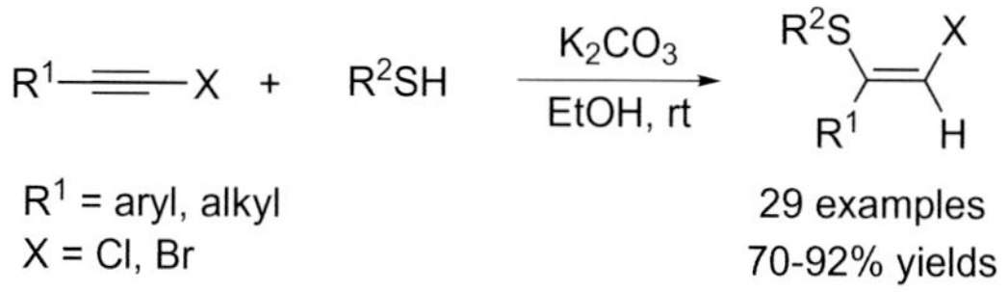

$R^1\!\!\equiv\!\!-X$ + R^2SH

$$\xrightarrow[\text{EtOH, rt}]{\text{K}_2\text{CO}_3}$$

R¹ = aryl, alkyl
X = Cl, Br

29 examples
70-92% yields

Scheme 3.45 K₂CO₃-promoted hydrothiolation of haloalkynes

3.2.1.6 Sulfur Nucleophiles

The importance of β-halo alkenyl sulfides has made them attract the attention of many scientists [188]. One of the most effective methods to access these compounds is the hydrothiolation of haloalkynes. In 2014, Zhu's group documented a K₂CO₃-promoted hydrothiolation reaction of haloalkynes, producing β-halo alkenyl sulfides in high yields with excellent regio- and stereoselectivities. This operationally simple and efficient protocol tolerated diverse functional groups (Scheme 3.45) [189].

General Procedure for K₂CO₃-Promoted Hydrothiolation of Haloalkynes: To a mixture of 2-mercaptopyridine (0.6 mmol) and K₂CO₃ (0.65 mmol) in EtOH (2 mL) was added haloalkyne (0.5 mmol). After stirring at room temperature for 10 h, the reaction mixture was quenched with water, extracted with EtOAc, dried over Na₂SO₄ and concentrated. Column chromatography on silica gel gave the β-halo alkenyl sulfide products.

3.2.2   Cycloadditions

Transition metal-catalyzed cycloadditions have demonstrated their great value in the efficient construction of ring systems and complex skeletons [190, 191]. Due to their electron-withdrawing properties, haloalkynes could potentially accelerate the reaction rate of cycloaddition. In this context, transition metal-catalyzed cycload-ditions of haloalkynes have attracted considerable attention. Additionally, the halide moiety could be utilized for further decoration, providing an alternative protocol for those cyclic structures difficult to access via direct cycloaddition procedure.

3.2.2.1   [2 + 2] Cycloaddition

The [2 + 2] cycloadditions between alkynes and alkenes are known to be an efficient method for the construction of cyclobutene rings [192, 193]. In 2004, Tam's group developed the [2 + 2] cycloaddition of bicyclic alkenes with haloalkynes under Ru catalysis (Scheme 3.46) [194]. Notably, the halide moiety greatly improved the reactivity of the alkyne component in the cycloaddition reaction. Importantly, the obtained cycloadducts could be transferred into various products via nucleophilic addition, Suzuki coupling, and Sonogashira coupling. Mechanism studies indicated that chloroalkynes reacted faster than bromoalkynes in this cycloaddition [195, 196]. Later, Koldobskii [197, 198] reported the [2 + 2] cycloaddition reaction of haloalkynes and vinyl ethers.

General Procedure for Ru-Catalyzed [2 + 2] Cycloaddition between Norbornadiene and Haloalkynes: A mixture of nobornadiene (3–5 equiv), and haloalkyne (1 equiv) in THF (0.5 mmol/mL) was added via a cannula to an oven-dried screw-cap vial containing Cp*RuCl(COD) (10 mol%) under nitrogen. The reaction mixture was stirred in the dark at 25–65 °C for 1–168 h. The crude product was purified by column chromatography to give the cycloadduct.

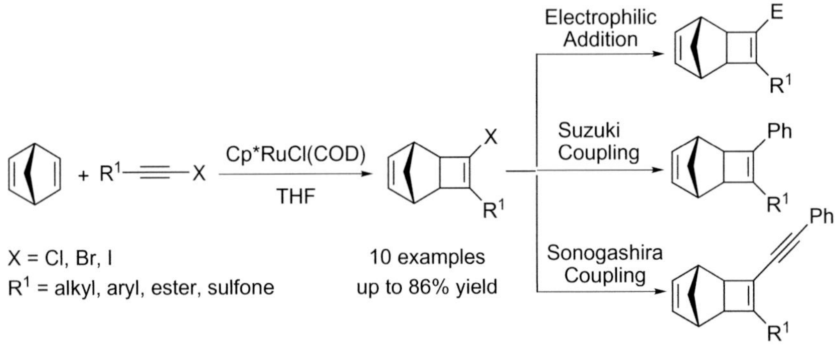

Scheme 3.46   Ru-catalyzed [2 + 2] cycloaddition between norbornadiene and haloalkynes

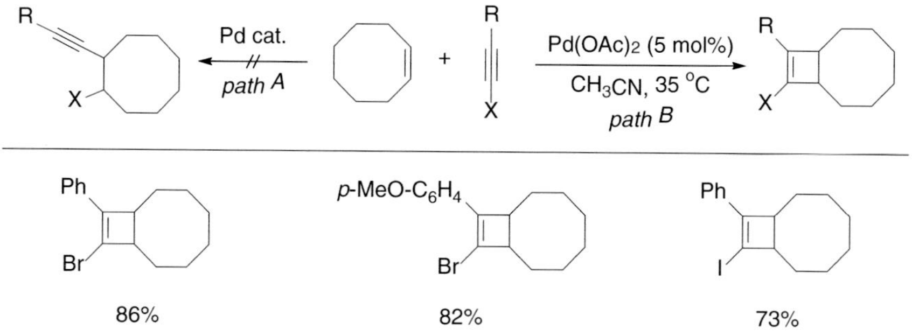

Scheme 3.47 Cycloaddition of haloalkynes and cyclooctene

In sharp contrast, the [2 + 2] cycloaddition of monocyclic alkenes with alkynes continues to represent a synthetic challenge. In 2011, Jiang and coworkers discovered that cyclooctene, a flexible alkene rather than the strained norbornene, reacted with haloalkyne could lead to a four-membered ring system via [2 + 2] cycloaddition pathway under mild conditions (Scheme 3.47, path B), while the 3-propynyl halide derivatives were not detected (Scheme 3.47, path A) [101]. This approach was another representative example of haloalkynes for carbocycle formation under Pd catalysis. Importantly, aromatic alkynyl bromides, with either electron-donating or electron-withdrawing groups attached to the benzene rings, were able to undergo the [2 + 2] cycloaddition smoothly and delivered the corresponding cycloadducts in moderate to good yields. However, cyclododecene was found to be completely ineffective under the optimized conditions, while cycloheptene afforded an inseparable mixture including the Alder-ene products. These observations indicated that the ring size of the cyclicalkene was crucial for the formation of the desired cyclobutene derivatives.

General Procedure for the Cycloaddition of Haloalkynes and Cyclooctene: To a Schlenk tube (25 mL) was successively added Pd(OAc)$_2$ (5 mol%), CH$_3$CN (2 mL), cyclooctene (1.3 mmol) and haloalkyne (1 mmol). The resulting mixture was stirred at 30 °C for 10 h. Then, the mixture was filtered through a small amount of silica gel. The filtrate was concentrated under reduced pressure and the residue was purified by silica gel preparative TLC (*n*-hexane) to give the desired product.

Additionally, a unique example was reported by Mikami in 2011 [199]. The catalytic asymmetric [2 + 2] cycloaddition reaction of 1-iodoalkyne with ethyl trifluoropyruvate was realized in the presence of a palladium catalyst **43**. Although the author only presented three substrates, this reaction indeed represented the first example of catalytic asymmetric [2 + 2] cycloaddition reaction of haloalkyne with a carbonyl group (Scheme 3.48).

General Procedure for Pd-Catalyzed Asymmetric [2 + 2] Cycloaddition Reaction of 1-Iodoalkyne: To a solution of (*S*)-BINAP-PdCl$_2$ (2 mol%) in CH$_2$Cl$_2$ (2 mL) was added AgSbF$_6$ (2.2 mol%) at room temperature under argon atmosphere. After stirring for 30 min, ethyl trifluoropyruvate (1 mmol) and iodoalkyne (0.5 mmol) were added to the mixture at −20 °C for 12 h, and then the reaction

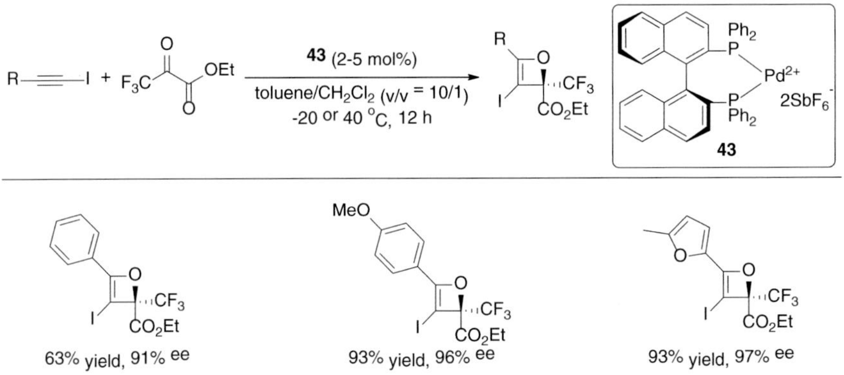

Scheme 3.48 Pd-catalyzed asymmetric [2 + 2] cycloaddition reaction of 1-iodoalkyne

mixture was directly loaded onto a short silica-gel column to remove the catalyst. Purification by silica-gel chromatography gave the corresponding oxetene product. And the enantiomeric excess was determined by chiral HPLC analysis.

3.2.2.2 [3 + 2] Cycloaddition

The Cu-catalyzed azide–alkyne [3 + 2] cycloaddition reaction has been widely investigated in the field of synthetic and medicinal chemistry, polymer chemistry, and materials science [200]. However, the efficiency and selectivity of this transformation depend on the reactivity of *in situ* generated Cu^I acetylides. Therefore the reaction partners are usually limited to terminal acetylenes, which provide only 1,4-disubstituted triazoles. In this regard, a general and practical protocol for the regio-controlled construction of different substituted triazoles would be a valuable complement to the "click chemistry". One outstanding example is that the efficient method reported by Hein and Fokin et al., for the chemo- and regioselective synthesis of iodotriazoles from organic azides and iodoalkynes (Scheme 3.49) [201]. This reaction featured a broad substrate scope, excellent functional group and solvent tolerance, and also remarkably high reaction rates. The employment of TTTA as ligand was the key to achieve this transformation, because no reaction was observed when TTTA was omitted, and the chemoselectivity as well as the observed rate of the reaction were strongly dependent on the nature of the ligand. As an additional benefit, the 5-iodo-1,2,3-triazole adducts are versatile synthetic intermediates, which are amenable to further functionalization. Later, García-Álvarez [202], Rowan [203], Zhu [204], and Díez-González [205] independently reported the cycloaddition of azides with haloalkynes under copper catalysis.

General Procedure for Copper-Catalyzed Cycloaddition of Azides with 1-Iodoalkynes: CuI (5 mol%) and TTTA (5 mol%) were stirred in THF (4.5 mL) at room temperature for 20 min, after which time a homogeneous solution was obtained. Organic azide (1 mmol) and 1-iodoalkyne (1 mmol) were dissolved in

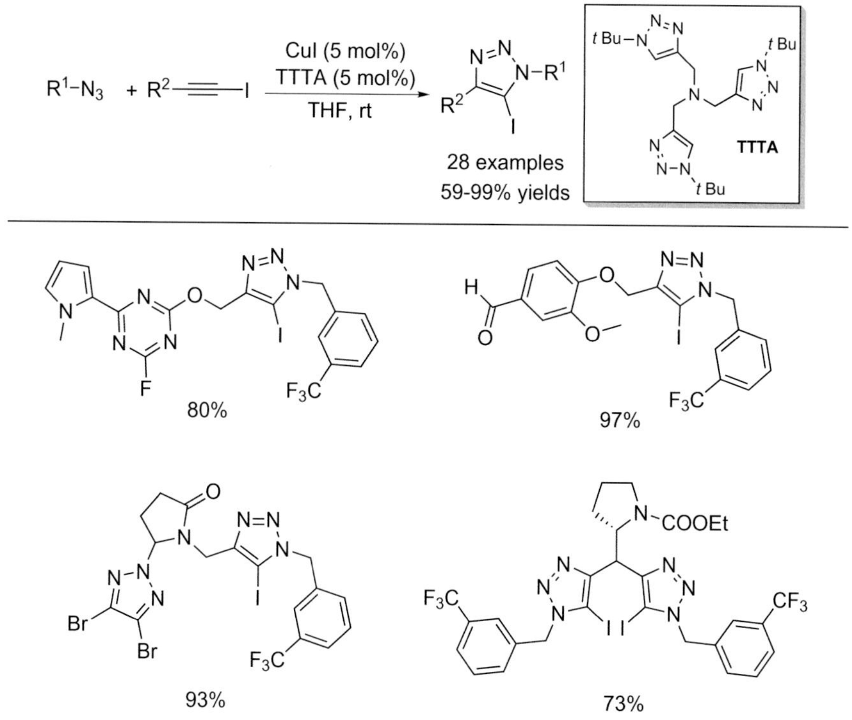

Scheme 3.49 Copper-catalyzed cycloaddition of azides with 1-iodoalkynes

THF (0.5 mL) and added in a single portion to the catalyst solution. The reaction mixture was stirred for 45 min, and then quenched by adding 10% NH_4OH solution (1 mL). The volatile components were removed by evaporation, and the resulting residue was suspended in water and diethyl ether. A precipitate formed upon vigorous stirring and was isolated by filtration to give the triazole as white powder.

Base on the experimental results and previous literatures [206, 207], the authors outlined their mechanistic proposals in Scheme 3.50. In path A, firstly, σ-acetylide complex **44** was formed. Then, key intermediate **44** coordinated to the proximal nitrogen center and subsequent cyclization to afford the cuprated triazole **45**. Finally, copper exchanged with iodoalkyne via σ-bond metathesis to provide the iodotriazole product and regenerate the acetylide **44**. On the other hand in path B, copper might activate the iodoalkyne through the formation of a π-complex intermediate **46**, which would engage the azide to deliver complex **47**. Then the complex **47** underwent cyclization through a vinyldiene-like transition state **48** to produce the triazole product.

The isoxazole moiety is also an attractive pharmacophoric element which is found in various useful therapeutic agents [208, 209]. The [3 + 2] cycloaddition of nitrile oxides with haloalkynes is an efficient route to facile access halo substituted isoxazole compounds, which could be further functionalized. Although this reaction

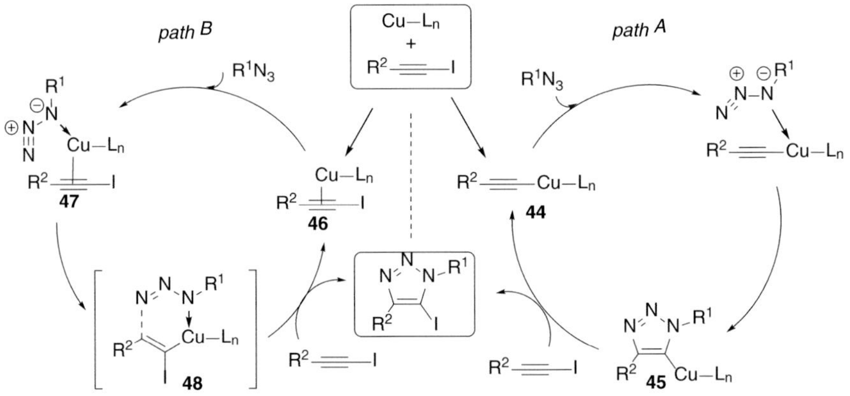

Scheme 3.50 Proposed mechanism for the copper-catalyzed azide-iodoalkyne cycloaddition

was investigated early in 1989, it suffered limited substrate scope or poor yields [210, 211]. In 2010, Browne and co-workers reported a thermally promoted cycloaddition of iodoalkynes with in situ generated nitrile oxides from chloro-oximes. This method has a broad substrate scope with respect to both iodoalkynes and chloro-oximes, and delivered the corresponding isoxazole products in good yields with excellent regioselectivity (Scheme 3.51) [212].

General Procedure for the Cycloaddition of 1-Iodoalkynes and Nitrile Oxides: The chloro-oxime (0.5 mmol), iodoalkyne (1 mmol), and DME (3 mL) were added to a two-necked round-bottom flask, which was then equipped with a suba seal and a condenser. The mixture was heated to 100 °C for 24 h with syringe pump addition of a Na_2CO_3 aqueous solution (2.1 mL, 0.25 M in water). Then the reaction was cooled, extracted with DCM, dried with $MgSO_4$, filtered, and concentrated. The residue was purified by column chromatography on silica gel.

Imidazo-containing motifs are versatile building blocks in natural products and bioactive compounds that have great significance in the area of pharmaceuticals [213, 214]. Undoubtedly, the intermolecular oxidative diamination of haloalkynes via a transition metal-catalyzed nucleophilic addition/C–N bond formation cascade process is an attractive approach to synthesize imidazo derivatives, in which the reactive halogen substituent of the haloalkynes was retained. In 2012, Jiang and co-workers revealed a new and direct approach to construct 2-halo-substituted imidazo[1,2-a]pyridines through the Cu-catalyzed oxidative cyclization reaction of o-aminopyridines and haloalkynes. Various 2-halo-substituted imidazopyridine, imidazopyrazine and imidazopyrimidine products were obtained with high regioselectivity under mild reaction conditions (Scheme 3.52) [215]. Furthermore, the resultant 2-halo-substituted products could be easily functionalized via elegant cross-coupling reactions. A highly conjugated structure **49** was successfully constructed after three-step synthesis. The practicality of this Cu-catalyzed oxidative cyclization reaction exhibited its potential utilities for the construction of optoelectronic materials (Scheme 3.53).

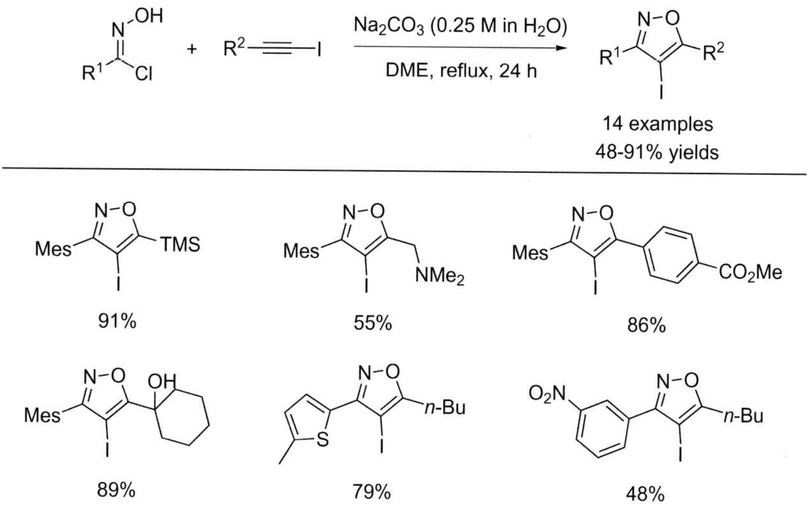

Scheme 3.51 The synthesis of isoxazoles

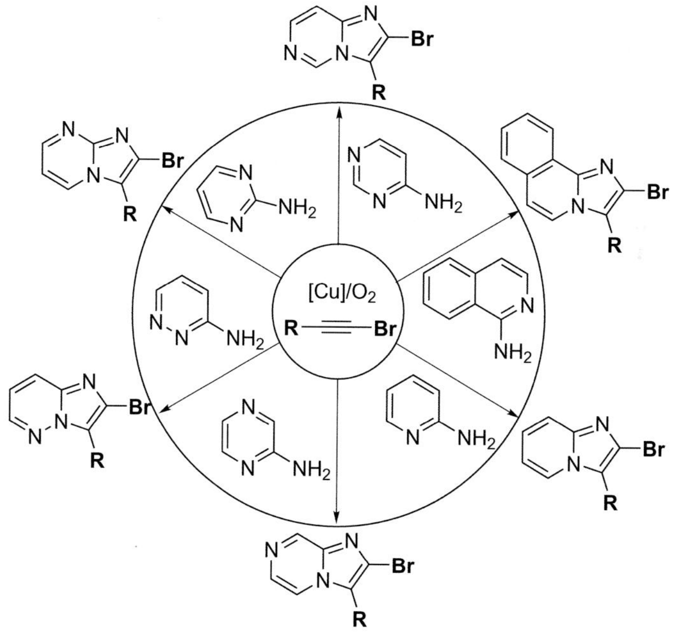

Scheme 3.52 Cu-catalyzed of synthesis of imidazopyridine structures

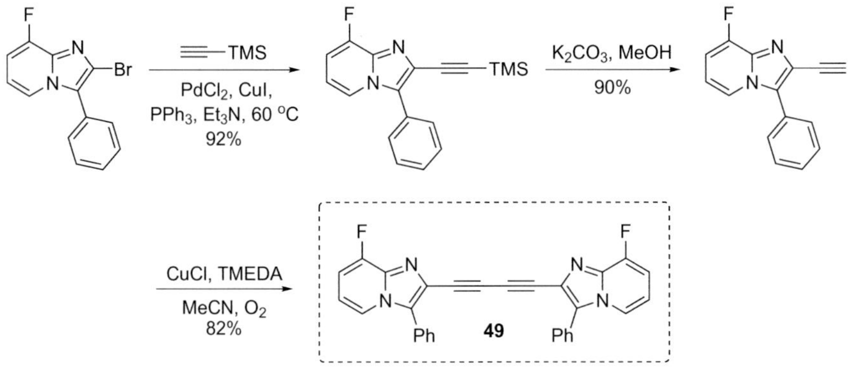

Scheme 3.53 Synthetic applications of imidazopyridine products

General Procedure for Cu-Catalyzed of Synthesis of Imidazopyridine Structures: A mixture of 2-aminopyridine (0.3 mmol), haloalkyne (0.2 mmol), and Cu(OTf)$_2$ (20 mol%) was stirred in MeCN (2 mL) at 60 °C under an oxygen atmosphere for 12 h. Then, water (10 mL) was added to quench the reaction. The aqueous solution was extracted with diethyl ether (10 mL × 3) and the combined organic layers were dried with MgSO$_4$, filtered and concentrated in vacuum. The residue was separated by flash column chromatography on silica gel to give the imidazopyridine products.

Pyrroles represent an interesting class of nitrogen-containing heterocycles that exhibit diverse therapeutic and biological activities [216–218]. Among them, 3-halo-substituted pyrroles are quite appealing as they provide a facile method for the deravatization at the 3-position of pyrroles. However, the examples for their efficient synthesis are still very rare. [219, 220]. In 2015, Jiang's group documented a novel palladium-catalyzed oxidative cyclization of bromoalkynes with N-alkylamines via cascade formation of C–N and C–C bond [221]. A wide spectrum of 3-bromopyrroles were obtained in moderate to excellent yields. Furthermore, the resultant 3-bromopyrroles could be easily functionalized via elegant cross-coupling reactions (Scheme 3.54).

General Procedure for Pd-Catalyzed of Synthesis of 3-Halo-Substituted Pyrroles: N-Allylamine (0.2 mmol), bromoalkynes (0.2 mmol), PdCl$_2$ (10 mol%) and BQ (2 equiv) were added to a solution of toluene/DMSO (2 mL, v/v = 5/1). The mixture was stirred under air at 110 °C. Upon completion, water (15 mL) was added and the resulting mixture was extracted with ethyl acetate (15 mL × 2). The combined organic phase was dried over Na$_2$SO$_4$, filtered and concentrated. The residue was eventually purified by flash column chromatography on silica gel with petroleum ether/ethyl acetate as eluent to afford the corresponding pyrroles.

On the basis of experimental data and previous reports [222, 223], a tentative reaction mechanism for this transformation was proposed in Scheme 3.55. Initially, the intermediate **50** was generated by the reaction of palladium(II) and N-allyamine. Subsequently, an intermolecular *cis*-insertion of bromoalkyne into the N-Pd bond gave the intermediate **51**, which underwent 1,2-migratory insertion, delivering the

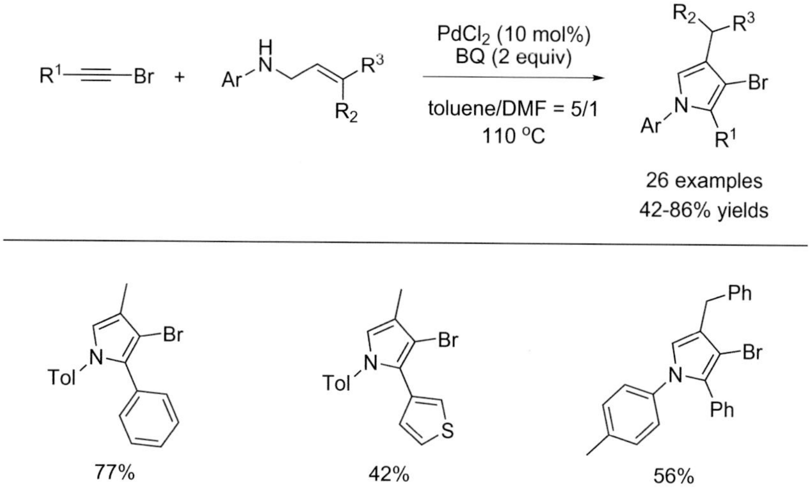

Scheme 3.54 Pd-catalyzed of synthesis of 3-halo-substituted pyrroles

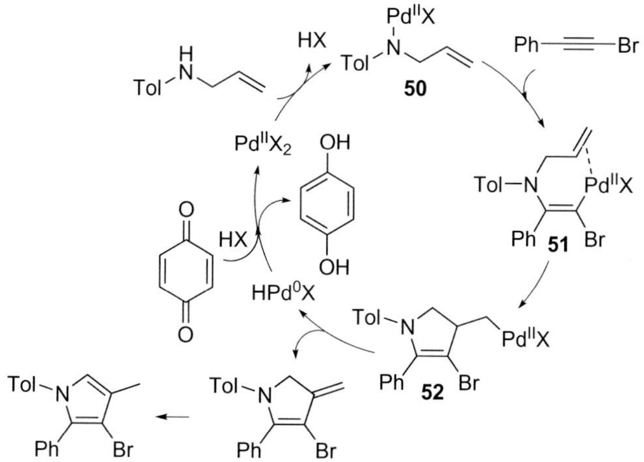

Scheme 3.55 Proposed mechanism

species **52**. A sequence of β-hydride elimination and isomerization afforded the desired 3-bromopyrrole adducts. In the meantime, palladium(0) species was reoxidized to palladium(II) species by BQ (1,4-benzoquinone).

3.2.2.3 [4 + 2] and [2 + 2 + 2] Cycloadditions

Transition metal-catalyzed [4 + 2] cycloaddition represents one of the most straightforward and efficient methods for the construction of six-membered rings [224, 225]. Due to the diverse transformation of the carbon carbon triple bond motif, haloalkynes have also exhibited their applications in transition metal-catalyzed [4 + 2] cycloaddition reactions [226]. In 2005, Tam's group [227] documented the first example of cationic Rh-catalyzed intramolecular [4 + 2] cycloaddition reaction of diene-tethered alkynyl halides (Scheme 3.56). The halide unit was found to be compatible under this catalytic system. Significantly, the halogen-containing cycloadducts could be converted into various products of synthetic usefulness.

General Procedure for Rh-Catalyzed [4 + 2] Cycloaddition of Diene-Tethered Alkynyl Halides: Inside an inert atmosphere (Ar) Glove Box, [RhCl(COD)]$_2$ (2.5 mol%) and AgSbF$_4$ (5 mol%) was added to an oven-dried vial and dissolved in acetone (3 mL). The reaction mixture was allowed to stir for 30 min and then added to another oven-dried vial containing the diene-tethered alkynyl halide (0.2 mmol) dissolved in acetone (6 mL). The reaction mixture was stirred at room temperature for 30 min. The crude reaction mixture was purified by column chromatography (EtOAc:hexanes = 1:9) to provide the desired product.

Additionally, another alternative method for the synthesis of six-membered rings is the [2 + 2 + 2] cycloaddition reaction [228]. In 2009, Nicolaou et al. [229] firstly reported the total synthesis of sporolides B, an unusual natural product isolated from the marine actinomycete *Salinospora tropica*. Crucially, they forged the chlorobenzenoid indane structural motif through a regio- and stereoselective Ru-catalyzed intermolecular [2 + 2 + 2] cycloaddition reaction between two acetylenic motifs, one of which bearing the chlorine residue (Scheme 3.57). This excellent work showed the prominent potential application of haloalkyne derivatives in the total synthesis of natural product.

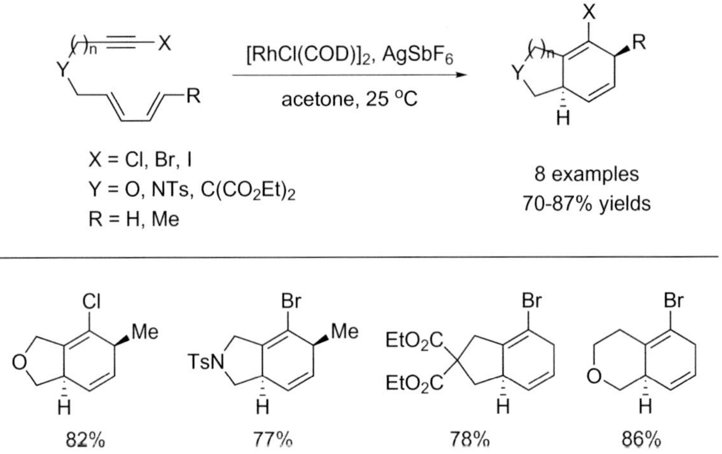

Scheme 3.56 Rh-catalyzed [4 + 2] cycloaddition of diene-tethered alkynyl halides

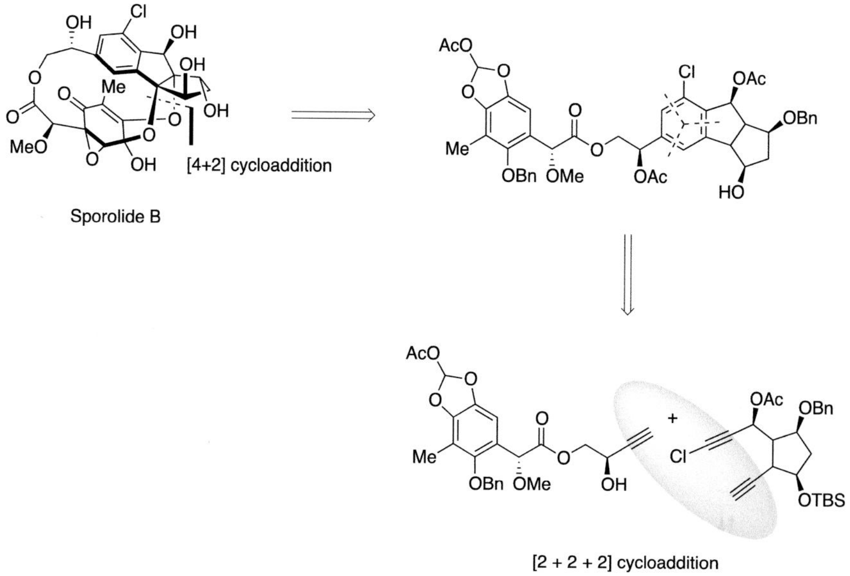

Sporolide B

[4+2] cycloaddition

[2 + 2 + 2] cycloaddition

Scheme 3.57 [2 + 2 + 2] Cycloaddition of haloalkyne in natural product synthesis

3.3   Transformations Involved Both Carbon-Halo Bond and Carbon-Carbon Triple Bond Motif

The development of efficient and practical methods for the construction of molecular complexity from simple and readily available reagents is an everlasting research topic in synthetic chemistry. The transformations of haloalkynes, involving both the carbon-carbon triple bond unit and the carbon-halo bond motif, have provided a valuable strategy to access various useful compounds.

3.3.1   Initially Reacted at the Carbon-Halo Bond

Unsaturated heterocyclic compounds are important synthetic intermediate as well as prevalent structural motifs found in natural and artificial molecules [230]. In 2008, Urabe and co-workers reported 1,2-double amination of haloalkynes via copper catalysis, a concise route for the synthesis of protected tetrahydropyrazines and related compounds. This reaction exhibited reasonable generality for aliphatic and aromatic haloalkynes, delivering the corresponding tetrahydropyrazines in good yields (Scheme 3.58) [231].

General Procedure for Copper-Catalyzed Diamination of Haloalkynes: To a mixture of *N,N'*-di(*p*-toluenesulfonyl)ethylenediamine (0.4 mmol), powdered K_3PO_4 (0.8 mmol), and CuI (5 mol%) was added haloalkyne (0.4 mmol) in DMF (4 mL), followed by *N,N'*-dimethylethylenediamine (0.1 mmol) under argon.

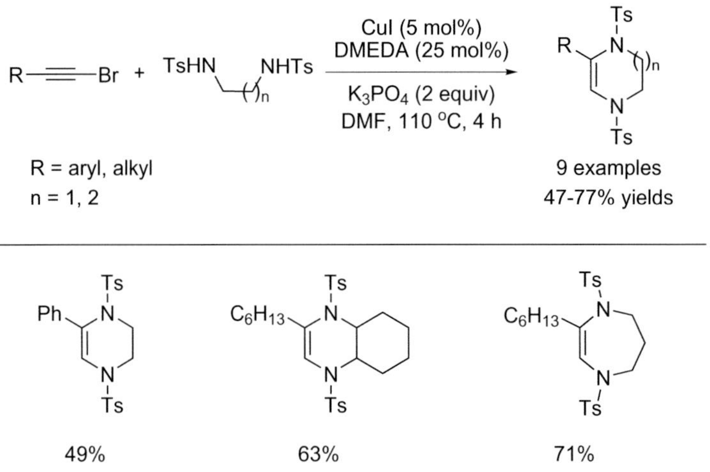

Scheme 3.58 Copper-catalyzed diamination of haloalkynes

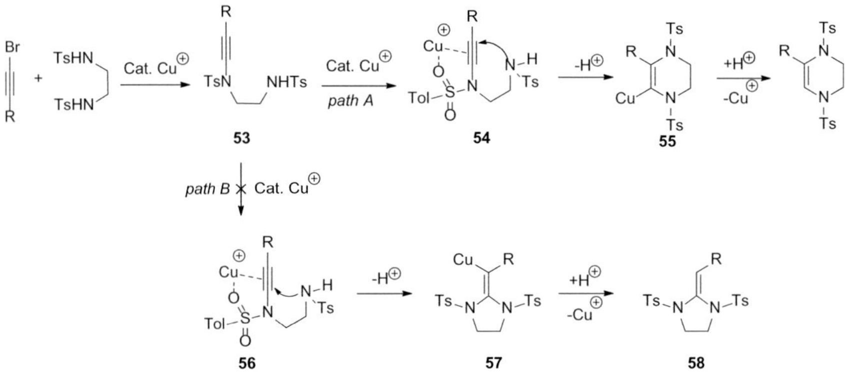

Scheme 3.59 Proposed reaction mechanism

The mixture was stirred in an oil bath maintained at 110 °C for 4 h. After being cooled to room temperature, the reaction mixture was diluted with water and extracted with ethyl acetate. The combined organic layers were dried over Na_2SO_4 and concentrated in vacuo to give a crude oil, which was purified by column chromatography on silica gel.

The author also proposed the reaction mechanism as outlined in Scheme 3.59. Firstly, the alkynylation of sulfonamide gave the ynamide intermediate 53, then the second amination of the acetylenic bond in 54 proceeded in a 6-endo-dig manner under copper catalysis to give cuprate 54. Finally, protonation of the intermediate 54 provided the observed product and released the copper catalyst (path A). Importantly, the formation of isomeric tetrahydroimidazole 58 via the cyclization of 5-exo-dig mode (56 to 57, path B) was not observed. Interestingly, when bromopropiolic acid derivatives were used, the 5-exo-dig type product could be obtained in good yield under transition-metal-free conditions [232]. Upon the diverse transformation abilities of ynamide, Jiang's group realized the synthesis of naphthalene-1,3-diamine derivatives from haloalkynes and amines under copper catalysis [233].

Amides are one of the most prevalent functional groups in natural products, pharmaceuticals, and polymers. In 2011, Jiang's group revealed a mild and efficient multi-component reaction for the construction of amides from bromoalkynes under transition-metal free conditions, which provided a wide range of secondary and tertiary amides in moderate to excellent yields (Scheme 3.60) [234]. The control experiments indicated that the alkynyl bromide should first react with amine to generate ynamine adduct and the isotopic labeling investigation clearly demonstrated that the oxygen atoms of the amide products originated from water [Scheme 3.60, Eqs. (1)–(3)]. Based on these observations, a mechanism involving ynamine intermediate formation and nucleophilic addition process was proposed. Later, they applied this method to construct thioamides [235].

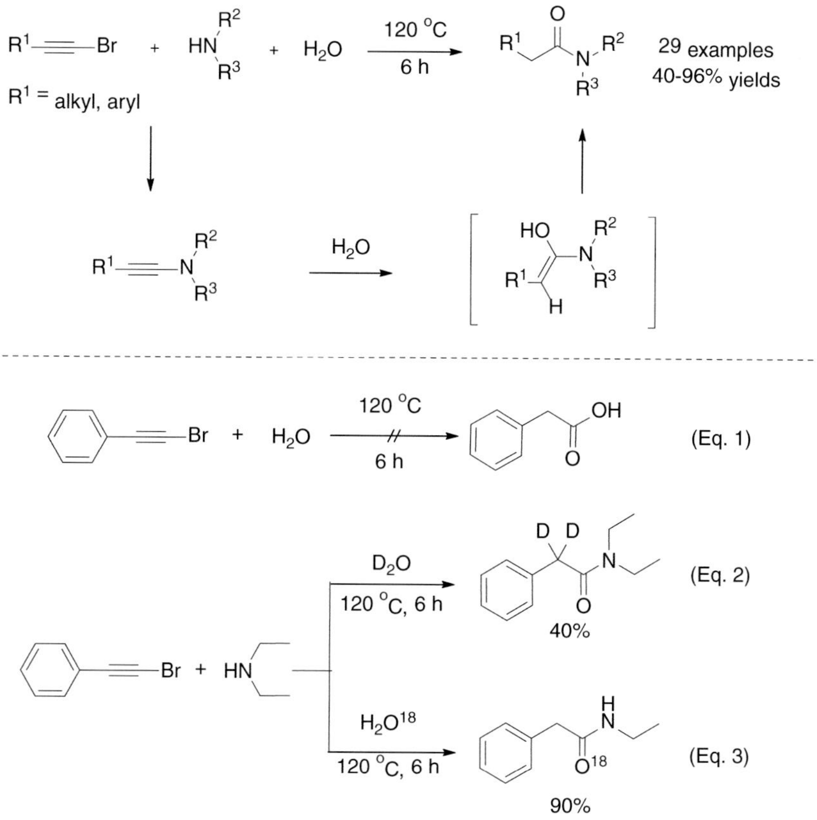

Scheme 3.60 Multi-component reaction for amides

General Procedure for the Synthesis of Amides: The mixture of 1-bromoalkyne (1 mmol) and amine (1.5 mmol) in water (2 mL) was stirred at 120 °C for 6 h in a Schlenk tube (25 mL). Upon completion of the reaction, water (8 mL) was added to the mixture. The resulting aqueous solution was extracted with diethyl ether (15 mL × 3). The combined organic phase was dried with anhydrous MgSO$_4$, filtered and concentrated. The residue was purified by column chromatography to give the corresponding amides.

3.3.2 Initially Reacted at the Carbon-Carbon Triple Bond

Benzo[*b*]furans, are versatile synthetic blocks and significant structural motifs of natural products and potential drugs [236–238]. Due to their potential applications, the development of practical and efficient synthetic methods is highly demanded. In 2011, Wang and co-workers reported a sequential, one-pot reaction of phenols with bromoalkynes for the synthesis of benzo[*b*]furans [239]. This reaction tolerated diverse functional groups, and delivered the corresponding benzo[*b*]furan products in good yields. Importantly, the reaction intermediate could be isolated, and gave the desired product in high yield under the standard reaction conditions (Scheme 3.61).

General Procedure for the Synthesis of Benzo[b]furans: Under air atmosphere, a sealable tube equipped with a magnetic stirrer bar was charged with phenol (1.1 mmol), bromoalkyne (1 mmol), K$_2$CO$_3$ (2 mmol) and DMF (2 mL). The rubber septum was then replaced with a Teflon-coated screw cap, and the reaction vessel was placed in an oil bath at 110 °C for 12 h. Then PdCl$_2$ (5 mol%) was added and the reaction was performed at 130 °C for 6 h. After the reaction was completed,

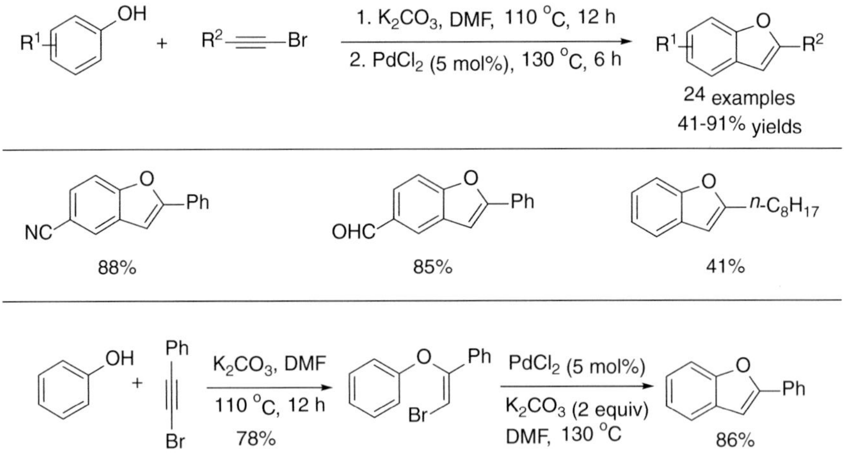

Scheme 3.61 The synthesis of benzo[*b*]furans

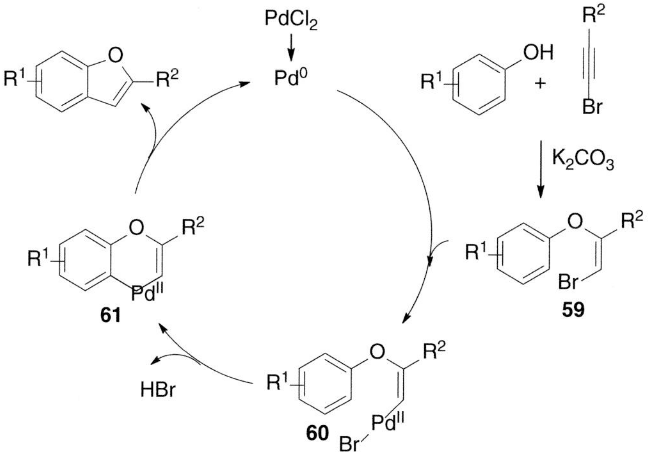

Scheme 3.62 Proposed mechanism

it was cooled to room temperature and diluted with ethyl acetate. The resulting solution was directly filtered through a pad of silica gel using a sintered glass funnel, and concentrated under reduced pressure. The residue was purified by chromatography on silica gel to give the benzo[*b*]furan product.

Based on these observations, the author proposed the possible reaction mechanism. Firstly, the intermolecular nucleophilic addition of phenol to bromoalkyne in the presence of base formed the (Z)-2-bromovinyl phenyl ether **59**. Then the intermediate **59** reacted with Pd^0 to give the Pd^{II} complex **60** via oxidative addition. Subsequently, an intramolecular electrophilic aromatic palladation of **60** generated the intermediate **61**, which was followed by a reductive elimination to provide the product and regenerate the Pd^0 catalyst (Scheme 3.62).

Except for furan derivatives [240], benzoxazepine derivatives, a very important kind of seven-membered ring, are the core building blocks with remarkable biological activities and pharmaceutical interests [241–243]. Unfortunately, multi-step synthesis is necessary for their preparation, which usually prevents them from constructing benzoxazepine analogues that are diverse in structure and electronic property. In 2012, Jiang and co-workers documented a robust route for the construction of substituted 4-amine-benzo[*b*] [1,4] oxazepines in a facile and convenient manner. This Pd-catalyzed cascade transformation of *o*-aminophenols, bromoalkynes and isocyanides underwent a selective C–O and C–N bond formation procedure and delivered the desired products in good to excellent yields (Scheme 3.63) [244].

General Procedure for Pd-Catalyzed Cascade Reaction for 4-Amine-benzo[b] [1,4] oxazepines: The mixture of 2-aminophenol (0.5 mmol), $Pd(PPh_3)_2Cl_2$ (5 mol%) and PPh_3 (10 mol%) in 1,4-dioxane (1 mL) was placed in a Schlenk tube. Then, Cs_2CO_3 (1 mmol) and bromoalkyne (0.5 mmol) were added successively. The mixture was stirred for five min at room temperature. Subsequently, isocyanide

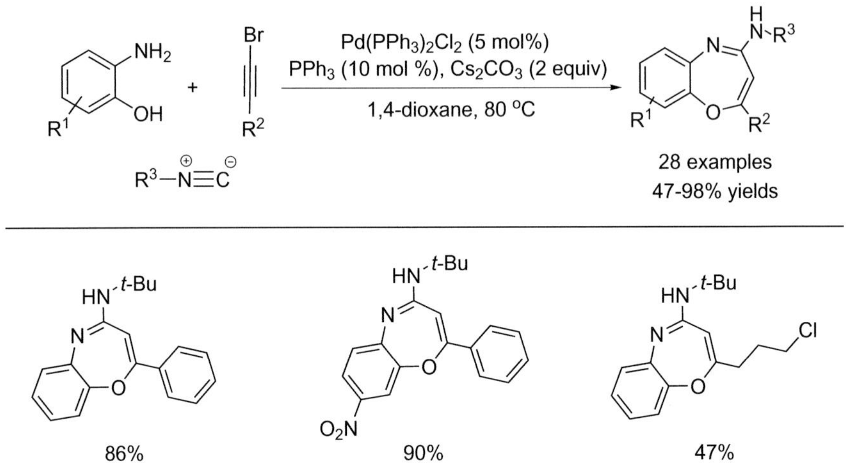

28 examples
47-98% yields

86%   90%   47%

Scheme 3.63 Pd-catalyzed cascade reaction for 4-amine-benzo[*b*][1,4]oxazepines

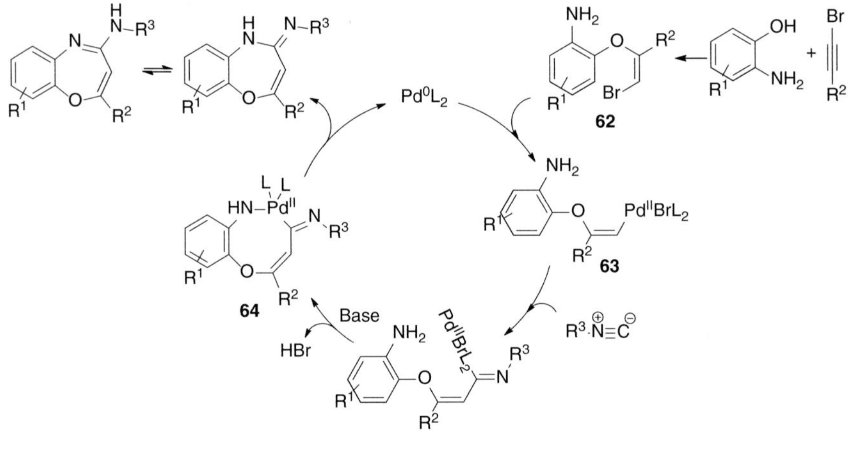

Scheme 3.64 Proposed mechanism

(0.6 mmol) was added in one portion. The resulting mixture was stirred at 80 °C for 2 h. Upon completion, the reaction mixture was extracted with ethyl acetate (10 mL × 3), and the combined organic layers were dried over anhydrous $MgSO_4$, filtered and concentrated under reduced pressure. The residue was purified by silica gel column chromatography (hexanes/EtOAc = 10/1) to give the corresponding product.

According to the mechanistic investigations, a possible catalytic cycle for this cascade reaction was illustrated in Scheme 3.64. Initially, nucleophilic addition of *o*-aminophenols to bromoalkynes delivered **62**, which underwent oxidative addition to Pd^0 species to form vinylpalladium(II) species **63**. Subsequently, migratory insertion of isocyanide and release of HBr under basic conditions gave the eight-membered azapalladacyclic intermediate **64**. Finally, reductive elimination and isomerization provided the benzoxazepine product, as well as regenerating the active Pd^0 catalyst.

Besides oxygen, nitrogen [245] nucleophiles triggered cascade annulation reactions of haloalkynes, carbon nucleophiles could also be used to construct cyclic compounds. In 2014, Jiang's group reported the first Pd-catalyzed annulation reaction of bromoalkynes and isocyanides, which provided a direct and practical route to a wide range of 5-iminopyrrolones with excellent reigoselectivity [Scheme 3.65, Eq. (1)] [172]. This reaction could proceed smoothly under mild reaction conditions, and broad functional groups could be tolerated. Intriguingly, they observed the formation of 2,5-diimino-furan as side product, an isomer with the same molecular weight as 5-iminopyrrolone [Scheme 3.65, Eq. (2)]. Systematically condition screening revealed that base and the reaction time were crucial for the reaction pathway. CsF and longer reaction time (8–12 h) preferred the formation of 5-iminopyrrolones, while K$_2$CO$_3$ and shorter reaction time (2 h) were favored to deliver the 2,5-diimino-furan products [246]. Furthermore, the resultant furans could readily undergo hydrolysis to give maleamide skeletons, which might have further applications in synthetic and medicinal chemistry.

General Procedure for Pd-Catalyzed Synthesis of 5-Iminopyrrolones: A mixture of Pd(OAc)$_2$ (5 mol%), H$_2$O (0.1 mL), DMSO (2 mL), isocyanide (3 mmol), haloalkyne (1 mmol) and CsF (1.5 mmol) was added successively in Schlenk tube. The mixture was stirred at 90 °C for 12 h. Upon completion of the reaction, the mixture was cooled to room temperature, and the solution was filtered through a small amount of silica gel. The solvent was removed under reduced pressure, and the residue was purified by silica gel preparative TLC to give the desired 5-iminopyrrolones product.

General Procedure for Pd-Catalyzed Synthesis of 2,5-Diimino-furans: A mixture of CsF (1.2 mmol), H$_2$O (0.1 mL), DMSO (2 mL), isocyanide (2 mmol), and bromoalkyne (1 mmol) was successively added in a Schlenk tube (25 mL). After

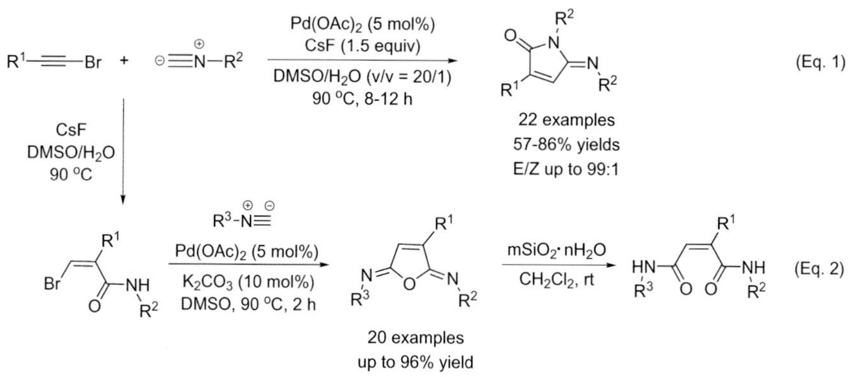

Scheme 3.65 Pd-catalyzed synthesis of 5-iminopyrrolones and 2,5-diimino-furans

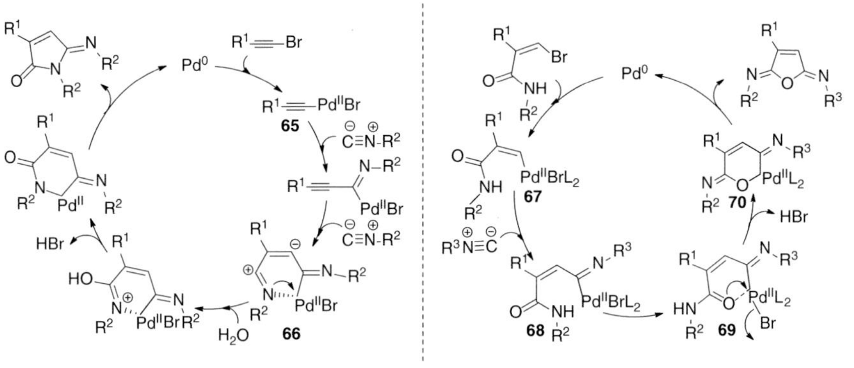

Scheme 3.66 Proposed mechanisms

stirring for 12 h at 90 °C, the starting materials were completely consumed as monitored by TLC and GC-MS analysis. Then, the reaction mixture was cooled to room temperature, filtered through a small amount of silica gel and concentrated. The residue was purified by silica gel preparative TLC to give the bromoacrylamide product.

A mixture of bromoacrylamide (0.2 mmol), isoacyanide (0.24 mmol), Pd(OAc)$_2$ (5 mol%), K$_2$CO$_3$ (0.4 mmol) and DMSO (2 mL) were added successively in a tube (10 mL). The resulting mixture was stirred at 90 °C for 2 h. Upon completion, the reaction mixture was extracted with ethyl acetate (10 mL × 3), and the organic layers were combined, dried over anhydrous MgSO$_4$, filtered and concentrated under reduced pressure. The residue was purified by aluminum oxide basic preparative TLC to give the 2,5-diimino-furan product.

Based on these findings, the authors proposed the mechanisms of the two reactions. For the synthesis of 5-iminopyrrolones (Scheme 3.66, left pathway), they believed the procedure was first initiated by the oxidative addition of Pd^0 species to bromoalkyne affording alkynylpalladium complex **65**, followed by migratory insertion and nucleophilic addition of isocyanide delivered intermediate **66**, in which the nitrogen atom would simultaneously coordinate with the Pd center. Then hydrolysis led to the release of HBr and 5-iminopyrrolone product was finally constructed by the reductive elimination with the regeneration of Pd^0 catalyst. As to the 2,5-diimino-furan derivatives (Scheme 3.66, right pathway), they proposed that the initial oxidative addition of Pd^0 to bromoacrylamide provided the vinylpalladium species **67**, subsequent migratory insertion of isocyanide generated **68**. Then, the coordination of the amide oxygen atom with the Pd^{II} center gave intermediate **69**. Finally, under the treatment of base, HBr would be eliminated to form complex **70**, which underwent the reductive elimination to deliver the annulation adduct and regenerated the active Pd^0 catalyst. It was supposed that the different coordinated type with Pd catalyst might be account for the one-pot reaction affording the N-containing cyclization products, whereas the two-step procedure giving the O-containing heterocycles.

References

1. Zhang B, Wang Y, Yang S, Zhou Y, Wu W, Tang W, Zuo J, Li Y, Yue J (2012) Ivorenolide A, an unprecedented immunosuppressive macrolide from Khaya ivorensis: structural elucidation and bioinspired total synthesis. J Am Chem Soc 134:20605–20608
2. Pålsson L, Wang C, Batsanov BS, King SM, Beeby A, Monkman AP, Bryce MR (2010) Efficient intramolecular charge transfer in oligoyne-linked donor-π-acceptor molecules. Chem Eur J 16:1470–1479
3. Glaser C (1870) Untersuchunge über einige Derivate der Zimmtsäure. Ann Chem Pharm 154:137–171
4. Eglinton G, Galbraith AR (1959) Macrocyclic acetylenic compounds. Part I. Cyclotetradeca-1:3-diyne and related compounds. J Chem Soc 889–896
5. Hay AS (1960) Communications-Oxidative coupling of acetylenes. J Org Chem 25:1725–1726
6. Hay AS (1962) Oxidative coupling of acetylenes. II[1]. J Org Chem 27:3320–3321
7. Rossi R, Carpita A, Bigelli C (1985) A palladium-promoted route to 3-alkyl-4-(1-alkynyl)-hexa-1,5-dyn-3-enes and/or 1,3-diynes. Tetrahedron Lett 26:523–526
8. Nishihara Y, Ikegashira K, Hirabayashi K, Ando J, Mori A, Hiyama T (2000) Coupling reactions of alkynylsilanes mediated by a Cu(I) salt: novel syntheses of conjugated diynes and disubstituted ethynes. J Org Chem 65:1780–1787
9. Damle SV, Seomoon D, Lee PH (2003) Palladium-catalyzed homocoupling reaction of 1-iodoalkynes: a simple and efficient synthesis of symmetrical 1,3-diynes. J Org Chem 68:7085–7087
10. Chen Z, Jiang H, Wang A, Yang S (2010) Transition-metal-free homocoupling of 1-haloalkynes: a facile synthesis of symmetrical 1,3-diynes. J Org Chem 75:6700–6703
11. Marino JP, Nguyen HN (2002) Bulky trialkysilyl acetylenes in the cadio-chodkiewicz cross-coupling reaction. J Org Chem 67:6841–6844
12. Jiang H, Wang A (2007) Copper-catalyzed cross-coupling reactions of bromoalkynols with terminal alkynes in supercritical carbon dioxide. Synthesis 11:1649–1654
13. Ahammed S, Kundu D, Ranu BC (2014) Cu-catalyzed Fe-driven Csp-Csp and Csp–Csp^2 cross-coupling: an access to 1,3-diynes and 1,3-enynes. J Org Chem 79:7391–7398
14. Negishi E, de Meijere A (2002) Handbook of organopalladium chemistry for organic synthesis. Wiley-Interscience, New York
15. Shi W, Luo Y, Luo X, Chao L, Zhang H, Wang J, Lei A (2008) Investigation of an efficient palladium-catalyzed C(sp)–C(sp) cross-coupling reaction using phosphine-olefin ligand: application and mechanistic aspects. J Am Chem Soc 130:14713–14720
16. Weng Y, Cheng B, He C, Lei A (2012) Rational design of a palladium-catalyzed Csp-Csp cross-coupling reaction inspired by kinetic studies. Angew Chem Int Ed 51:9547–9551
17. El-Jaber N, Estevez-Braun A, Ravelo AG, Munoz-Munoz O (2003) Acetylenic acid from the aerial parts of Nanodea muscosa. J Nat Prod 66:722–724
18. Miyaura N, Yamada K, Suginome H, Suzuki A (1985) Novel and convenient method for the stereo- and regiospecific synthesis of conjugated alkadienes and alkenynes via the palladium-catalyzed cross-coupling reaction of 1-alkenylboranes with bromoalkenes and bromoalkynes. J Am Chem Soc 107:972–980
19. Shi Y, Li X, Liu J, Jiang W, Sun L (2010) PdCl$_2$-catalyzed cross-coupling reaction of arylacetylene iodides with arylboronic acids to diarylacetylenes. Tetrahedron Lett 51:3626–3628
20. Wang S, Wang M, Wang L, Wang B, Li P, Yang J (2011) CuI-catalyzed Suzuki coupling reaction of organoboronic acids with alkynyl bromides. Tetrahedron 67:4800–4806
21. Shi Y, Li X, Liu J, Jiang W, Sun L (2011) A Suzuki-type cross-coupling reaction of arylacetylene halides with arylboronic acids. Appl Organomet Chem 25:514–520
22. Corpet M, Bai X, Gosmini C (2014) Cobalt-catalyzed cross-coupling of organozinc halides with bromoalkynes. Adv Synth Catal 356:2937–2942

23. Sämann C, Schade MA, Yamada S, Knochel P (2013) Functioinalized alkenylzinc reagents bearing carbonyl groups: preparation by direct insertion and reaction with electrophiles. Angew Chem Int Ed 52:9495–9499
24. Ma Y, Huang X (1997) A novel synthesis of 1,3-enynylselenides via cross coupling of (E)-α-selanylvinyl stannanes with haloalkynes. Synth Commun 29:3441–3447
25. Zhao H, Zhang H, Cai M (2008) A one-pot, stereoselective synthesis of (Z)-2-sulfonyl-substituted 1,3-enynes by hydrostannylation-stille tandem reaction of acetylenic sulfones. Chin J Chem 26:799–803
26. Jeon JH, Kim JH, Jeong YJ, Jeong IH (2014) Preparation of 2,2-difluoro-1-trialkylsilylethenylstannanes and their cross-coupling reactions. Tetrahedron Lett 55:1292–1295
27. Li Q, Ding Y, Yang X (2014) Nickel-catalyzed cross-coupling reaction of alkynyl bromides with Grignard reagents. Chin Chem Lett 25:1296–1300
28. Xie M, Huang X (2003) Preparation of selanyl-substituted conjugated enynes by copper-catalysed coupling reaction of (E)-γ-selanyl vinylzirconocene with acetylenic bromide. J Chem Res 584–585
29. Liu Y, Gao H (2006) New zirconium-mediated approach toward regio- and stereocontrolled synthesis of *trans*-enediynes. Org Lett 8:309–311
30. Cornelissen L, Lefrancq M, Riant O (2014) Copper-catalyzed cross-coupling of vinylsiloxanes with bromoalkynes: synthesis of enynes. Org Lett 16:3024–3027
31. Sonogashira K (2002) Development of Pd-Cu catalyzed cross-coupling of terminal acetylenes with sp^2-carbon halides. J Organomet Chem 653:46–49
32. Trofimov BA, Stepanova ZV, Sobenina LN, Mikhaleva AI, Ushakov IA (2004) Ethynylation of pyrroles with 1-acyl-2-bromoacetylenes on alumina: a formal 'Inverse Sonogashira Coupling'. Tetrahedron Lett 45:6513–6516
33. Dudnik AS, Gevorgyan V (2010) Formal inverse sonogashira reaction: direct alkynylation of arenes and heterocycles with alkynyl halides. Angew Chem Int Ed 49:2096–2098
34. Kalinin VK, Pashchenko DN, She FM (1992) Palladium-catalysed synthesis of 4-heteroaryl and 4-alkynyl-substituted sydnones. 5-oxido-3-phenyl-1,2,3-oxadiazol-3-ium-4-ylzinc chloride. Mendeleev Commun 2:60–61
35. Trofimov BA, Stepanova ZV, Sobenina LN, Mikhaleva AI, Ushakov IA (2004) Ethynylation of pyrroles with 1-acyl-2-bromoacetylenes on alumina: a formal 'inverse Sonogashira coupling'. Tetrahedron Lett 45:6513–6516
36. Sobenina LN, Demenev AP, Mikhaleva AI, Ushskov IA, Vasil'tsov AM, Ivanov AV, Trofimov BA (2006) Ethynylation of indoles with 1-benzoyl-2-bromoacetylene on Al$_2$O$_3$. Tetrahedron Lett 47:7139−7141
37. Trofimov BA, Sobenina LN, Demenev AP, Stepanova ZV, Petrova OV, Ushakov IA, Mikhaleva AI (2007) A palladium- and copper-free cross-coupling of ethyl-3-halo-2-propynoates with 4,5,6,7-tetrahydroindoles on alumina. Tetrahedron 48:4661–4664
38. Trofimov BA, Sobenina LN, Stepanova ZV, Petrova OV, Ushakov IA, Mikhaleva AI (2008) Chemo- and regioselective ethynylation of 4,5,6,7-tetrahydroindoles with ethyl 3-halo-2-propynoates. Tetrahedron Lett 49:3946–3949
39. Sobenina LN, Tomilin DN, Petrova OV, Gulia N, Osowska K, Szafert S, Mikhaleva AI, Trofimov BA (2010) Cross-coupling of 4,5,6,7-tetrahydroindole with functionalized haloacetylenes on active surfaces of metal oxides and salts. Russ J Org Chem 46:1373–1377
40. Sobenina LN, Stepanova ZV, Petrova OV, Ma JS, Yang G, Tatarinova AA, Mikhaleva AI, Trofimov BA (2013) Synthesis of 3-[5-(biphenyl-4-yl)pyrrol-2-yl]-1-phenylprop-2-yn-1-ones by palladium-free cross-coupling between pyrroles and haloalkynes on aluminum oxide. Russ Chem Bull Int Ed 62:88–92
41. Sobenina LN, Tomilin DN, Gotsko MD, Ushakov IA, Mikhaleva AI, Trofimov BA (2014) From 4,5,6,7-tetrahydroindoles to 3- or 5-(4,5,6,7-tetrahydroindole-2-yl)isoxazoles in two steps: a regioselective switch between 3- and 5-isomers. Tetrahedron 70:5168–5174
42. Sobenina LN, Petrova OV, Tomilin DN, Gotsko MD, Ushakov IA, Klvba LV, Mikhaleva AI, Trofimov BA (2014) Ethynylation of 2-(furan-2-yl)- and 2-(thiophen-2-yl)pyrroles with

acylbromoacetylenes in the Al$_2$O$_3$ medium: relative reactivity of heterocycles. Tetrahedron 70:9506–9511

43. Seregin IV, Ryabova V, Gevorgyan V (2007) Direct palladium-catalyzed alkynylation of N-fused heterocycles. J Am Chem Soc 129:7742–7743

44. Gu Y, Wang X (2009) Direct palladium-catalyzed C-3 alkynylation of indoles. Tetrahedron Lett 50:763–766

45. Kim SH, Chang S (2011) Highly efficient and versatile Pd-catalyzed direct alkynylation of both azoles and azolines. Org Lett 12:1868–1871

46. Wen Y, Wang A, Jiang H, Zhu S, Huang L (2011) Highly regio- and stereoselective synthesis of 1,3-enynes from unactivated ethylenes via palladium-catalyzed cross-coupling. Tetrahedron Lett 52:5736–5739

47. Wen Y, Jiang H (2012) Palladium-catalyzed coupling reaction of 1-bromoalkynes with olefins to synthesize enynes. Acta Chim Sinica 70:1716–1720

48. Xu Y, Zhang Q, He T, Meng F, Loh T (2014) Palladium-catalyzed direct alkynylation of N-vinylacetamides. Adv Synth Catal 356:1539–1543

49. Besselièvre F, Piguel S (2009) Copper as a powerful catalyst in the direct alkynylation of azoles. Angew Chem Int Ed 48:9553–9556

50. Phipps RJ, Guant MJ (2009) A meta-selective copper-catalyzed C−H bond arylation. Science 323:1593–1597

51. Streter ER, Bhayana B, Buchwald SL (2009) Mechanistic studies on the copper-catalyzed N-arylation of amides. J Am Chem Soc 131:78–88

52. Kawano T, Mstsuyama N, Hirano K, Satoh T, Miura M (2009) Room temperature direct alkynylation of 1,3,4-oxadiazoles with alkynyl bromides under copper catalysis. J Org Chem 75:1764–1766

53. Reddy GC, Balasubramanyam P, Salvanna N, Das B (2012) Copper-mediated C−H activation of 1,3,4-oxadiazoles with 1,1-dibromo-1-alkenes using PEG-400 as a solvent medium: distinct approach for the alkynylation of 1,3,4-oxadiazoles. Eur J Org Chem 2012:471–474

54. Matsuyama N, Hirano K, Satoh T, Miura M (2009) Nickel-catalyzed direct alkynylation of azoles with alkynyl bromides. Org Lett 11:4156–4159

55. Cacchi S, Fabrizi G, Moro L (1998) 2-substituted-3-allenyl-benzo[b]furans through the palladium-catalysed cyclization of propargylic o-(alkynyl)phenyl ethers. Tetrahedron Lett 39:5101–5104

56. Dai G, Larock RC (2002) Synthesis of 3,4-disubstituted Isoquinolines via palladium-catalyzed cross-coupling of 2-(1-alkynyl)benzaldimines and organic halides. J Org Chem 68:920–928

57. Arcadi A, Cacchi S, Fabrizi G, Marinelli F, Parisi LM (2005) Palladium-catalyzed reaction of o-alkynyltrifluoroacetanilides with 1-bromoalkynes. An approach to 2-substituted 3-alkynylindoles and 2-substituted 3-acylindoles. J Org Chem 70:6213–6217

58. Fujino D, Horimitsu H, Osuka A (2012) Synthesis of 1,2-disubstituted cyclopentenes by palladium-catalyzed reaction of homopropargyl-substituted dicarbonyl compounds with organic halides via 5-endo-dig cyclization. Org Lett 14:2914–2917

59. Kobayshi K, Arisawa M, Yamaguchi M (2002) GaCl$_3$-catalyzed ortho-ethynylation of phenols. J Am Chem Soc 124:8528–8529

60. Amemiya R, Fujii A, Arisawa M, Yamaguchi M (2003) GaCl$_3$-catalyzed α-ethynylation reaction of silyl enol ethers. J Organomet Chem 686:94–100

61. Amemiya R, Fujii A, Yamaguchi M (2004) GaCl$_3$-catalyzed ortho-ethynylation reaction of N-benzylanilines. Tetrahedron Lett 45:4333–4335

62. Tobisu M, Ano Y, Chatani N (2009) Palladium-catalyzed direct alkynylation of C−H bonds in benzenes. Org Lett 11:3250–3252

63. Ano Y, Tobisu M, Chatani N (2012) Palladium-catalyzed direct ortho-alkynylation of aromatic carboxylic acid derivatives. Org Lett 14:354–357

64. Zhao Y, He G, Nack WA, Chen G (2012) Palladium-catalyzed alkenylation and alkynylation of ortho-C(sp^2)−H bonds of benzylamine picolinamides. Org Lett 14:2948–2951

65. Shirakawa E, Yoshida H, Kurahashi T, Nakao Y, Hiyama T (1998) Carbostannylation of alkynes catalyzed by an iminophosphine-palladium complex. J Am Chem Soc 120:2975–2976
66. Liu H, Zhong Z, Nakajima K, Takahashi T (2002) Alkynylzirconation of alkynes and application to one-pot bisalkynylation of alkynes. J Org Chem 67:7451–7456
67. Liu Y, Song F, Cong L (2005) A facile Zr-mediated approach to (Z)-enynols and its application to regio- and stereoselective synthesis of fully substituted dihydrofuranes. J Org Chem 70:6999–7002
68. Suginome M, Shirakura M, Yamamoto A (2006) Nickel-catalyzed addition of alkynylboranes to alkynes. J Am Chem Soc 128:14438–14439
69. Nakao Y, Hirata Y, Tanaka M, Hiyama T (2007) Nickel/BPh$_3$-catalyzed alkynylcyanation of alkynes and 1,2-dienes: an effeient route to highly functionalized conjugated enynes. Angew Chem Int Ed 47:385–387
70. Li Y, Liu X, Jiang H, Feng Z (2010) Expedient synthesis of functionalized conjugated enynes: palladium-catalyzed bromoalkynylation of alkynes. Angew Chem Int Ed 49:3338–3341
71. Faust R, Göbelt B, Weber C, Krieger C, Gross M, Gisselbrecht JP, Boudon C (1999) 2,3-Dialkynyl-1,4-diazabuta-1,3-dienes as novel π-systems: synthesis, structure, and electronic properties. Eur J Org Chem 1999:205–214
72. Yoshida H, Asatu Y, Mimura Y, Ito Y, Ohshita J, Takaki K (2011) Three-component coupling of arynes and organic bromides. Angew Chem Int Ed 50:9676–9679
73. Amemiya R, Yamaguchi M (2007) Gallium trichloride-catalyzed exhaustive α-ethylation reaction of 1-silylacetylenes. Adv Synth Catal 349:1011–1014
74. Lubin-Germain N, Hallonet A, Huguenot F, Palmier S, Uziel J, Augé J (2007) Ferrier-type alkynylation reaction mediated by indium. Org Lett 9:3679–3682
75. Lubin-Germain N, Baltaze J, Coste A, Hallonet A, Lauréano H, Legrave G, Uziel J, Augé J (2008) Direct C-glycosylation by indium-mediated alkynylation on sugar anomeric position. Org Lett 10:725–728
76. Youcef RA, Santos MD, Roussel S, Baltaze J, Lubin-Germain N, Uziel J (2009) Huigen cycloaddition reaction of C-alkynyl ribosides under micellar catalysis: synthesis of ribavirin analogues. J Org Chem 74:4318–4323
77. Nicolai S, Waser J (2011) Pd(0)-catalyzed oxy- and aminoalkynylation of olefins for the synthesis of tetrahydrofurans and pyrrolidines. Org Lett 13:6324–6327
78. Sun N, Li Y, Yin G, Jiang S (2013) Palladium-catalyzed alkynylative lactonization of unsaturated bicyclic carboxylic acids: synthesis of fused polycyclic γ-lactone compounds. Eur J Org Chem 2013:2541–2544
79. Gutekunst WR, Baran PS (2014) Applications of C−H functionalization logic to cyclobutane synthesis. J Org Chem 79:2430–2452
80. Lhermitte F, Knochel P (1998) Stereoselective allylic C−H activation with tertiary alkylboranes: a new method for preparing cycloalkyl derivatives with three adjacent stereocenters. Angew Chem Int Ed 37:2460–2461
81. Hupe E, Knochel P (2001) Formal enantioselective michael addition with umpolung of reactivity. Angew Chem Int Ed 40:3022–3024
82. Chen S, Williams RM (2006) Syntheses of highly functionalized δ, γ-unsaturated-α-amino acids. Tetrahedron 62:11572–11579
83. Zhang X, Burton DJ (2000) An alternative route for the preparation of α, α-difluoropropargylphosphonates. Tetrahedron Lett 41:7791–7794
84. Thaler T, Guo L, Mayer P, Knochel P (2011) Highly diastereoselective C(sp^3)-C(sp) cross-coupling reactions between 1,3- and 1,4-substituted cyclohexylzinc reagents and bromoalkynes through remote stereocontrol. Angew Chem Int Ed 50:2174–2177
85. Sämann C, Dhayalan V, Schreiner PR, Knochel P (2014) Synthesis of substitued adamantylzinc reagents using a Mg-insertion in the presence of ZnI$_2$ and further functionalizations. Org Lett 16:2418–2421

86. Millet A, Dailler D, Larini P, Baudoin O (2014) Ligand-controlled α- and β-arylation of acyclic N-Boc amines. Angew Chem Int Ed 53:2678–2682
87. Black HK, Hom DHS, Weedon BCL (1954) Studies with acetylenes. Part I. Some reactions of Grignard reagents with alk-1-ynyl and alk-1-enyl halides. J Chem Soc 1704−1709
88. Cahiez G, Gager O, Buendia J (2010) Copper-catalyzed cross-coupling of alkyl and aryl Grignard reagents with alkynyl halides. Angew Chem Int Ed 49:1278–1281
89. Normant JF, Cahiez G, Chuit C, Villeras J (1974) Reactivite des vinylcuivres. Application a la synthese d'alcools allyliques Substitues stereospecifiquement. Tetrahedron Lett 14:2407–2408
90. Normant JF, Cahiez G, Chuit C (1974) Organocuivreux vinyliques: II. Deuterolyse, iodolyse, couplage et alcoylation stereospecifiques des vinyl-cuivres. J Orgnomet Chem 77:269−279
91. Liu Y, Xi C, Hara R, Nakajima K, Yamazaki A, Kotora M, Takahashi T (2000) Preparation of diynes via selective bisalkynylation of zirconacycles. J Org Chem 65:6951–6957
92. Poulsen TB, Barnardi L, Aleman J, Overgaard J, Jørgensen KA (2007) Organocatalytic asymmetric direct α-alkynylation of cyclic β-ketoesters. J Am Chem Soc 129:441–449
93. Daugulis O, Do H, Shabashov D (2009) Palladium- and copper-catalyzed arylation of carbon-hydrogen bonds. Acc Chem Res 42:1074–1086
94. Jazzar R, Hitce J, Renaudat A, Sofack-Kreutzer J, Baudoin O (2010) Functionalization of organic molecules by transition-metal-catalyzed C(sp^3)-H activation. Chem Eur J 16:2654–2672
95. Lyons TW, Sanford MS (2010) Palladium-catalyzed ligand-directed C−H functionalization reaction. Chem Rev 110:1147–1169
96. Ano Y, Tobisu M, Chatani N (2011) Palladium-catalyzed direct ethynylation of C(sp^3)-H bonds in aliphatic carboxylic acid derivatives. J Am Chem Soc 133:12984–12986
97. He J, Wasa M, Chan KSL, Yu J (2013) Palladium(0)-catalyzed alynylation of C(sp^3)-H bonds. J Am Chem Soc 135:3387–3390
98. Chen Z, Duan X, Zhou P, Ali S, Luo J, Liang Y (2012) Palladium-catalyzed divergent reactions of α-diazocarbonyl compounds with allyic esters: construction of quaternary carbon centers. Angew Chem Int Ed 51:1370–1374
99. Ziadi A, Correa A, Martin R (2013) Formal γ-alkynylation of ketones via Pd-catalyzed C−C cleavage. Chem Commun 49:4286–4288
100. Feng Y, Xu Z, Mao L, Zhang F, Xu H (2013) Copper catalyzed decarboxylative alkynylation of quaternary α-cyano acetate salts. Org Lett 15:1472–1475
101. Li Y, Liu X, Jiang H, Liu B, Chen Z, Zhou P (2011) Palladium-catalyzed bromoalkynylation of C-C double bonds: ring-structure triggered synthesis of 7-alkynyl norbornanes and cyclobutenyl halides. Angew Chem Int Ed 50:6341–6345
102. Furuya T, Ritter T (2008) Carbon-fluorine reductive elimination from a high-valent palladium fluoride. J Am Chem Soc 130:10060–10061
103. Gericke KM, Chai DI, Bieler N, Lautens M (2009) The norbornene shuttle: multicomponent domino synthesis of tetrasubstituted helical alkenes through multiple C−H functionalization. Angew Chem Int Ed 48:1447–1451
104. Liu H, Chen C, Wang L, Tong X (2011) Pd(0)-catalyzed iodoalkynation of norbornene scaffolds: the remarkable solvent effect on reaction pathway. Org Lett 13:5072–5079
105. Li J, Kaoud TS, Laroche C, Dalby KN, Kerwin SM (2009) Synthesis and biological evaluation of p38α kinase-targeting dialkynylimidazoles. Bioorg Med Chem Lett 19:6293–6297
106. Witulski B, Schweikert T, Schollmeyer D, Nemkovich NA (2010) Synthesis and molecular properties of donor-π-spacer-acceptor ynamides with up to four conjugated alkyne units. Chem Commun 46:2953–2955
107. Lam TY, Wang Y, Danheiser RL (2013) Benzannulation via the reaction of yanmides and vinylketenes application to the synthesis of highly substituted indoles. J Org Chem 78:9396–9414
108. Zificsak CA, Mulder JA, Hsung RP, Rameshkumar C, Wei L (2001) Recent advances in the chemistry of ynamines and ynamides. Tetrahedron 57:7575–7606

109. Hsung RP (2006) Chemistry of electron-deficient ynamines and ynamides. Tetrahedron 62:3771–3775
110. Wang X, Yeom H, Fang L, He S, Ma Z (2014) Ynamides in ring forming transformations. Acc Chem Res 47:1736–1748
111. DeKorver KA, Li H, Lohse AG, Hayashi R, Lu Z, Zhang Y, Hsung RP (2010) Ynamides: a modern functional group for the new millennium. Chem Rev 110:5064–5106
112. Janousek Z, Collard J, Viehe HG (1972) Reaction of secondary acetamides with N-dichloromethylene-N, N-dimethylammonium chloride. Angew Chem Int Ed Engl 11:917–918
113. Balsamo A, Macchia B, Macchia F, Rossello A, Domiano P (1985) 3-(4-iodomethyl-2-oxo-1-azetidinyl)propynoic acid t-butyl ester: a new β-lactam derivative, synthesis of an ynamide by reaction of 4-iodomethylazetidin-2-one with the t-butyl ester of propiolic acid in the presence of copper(I). Tetrahedron Lett 26:4141–4144
114. Balsells J, Vázquez J, Moyano A, Pericàs MA, Riera A (2000) Low-energy pathway for pauson-khand reactions: synthesis and reactivity of dicobalt hexacarbonyl complexes of chiral ynamines. J Org Chem 65:7291–7302
115. Frederick MO, Mulder JA, Tracey MR, Hsung RP, Huang J, Kurtz KCM, Shen L, Douglas CJ (2003) A copper-catalyzed C−N bond formation involving sp-hybridized carbons. A direct entry to chiral ynamides via N-alkynylation of amides. J Am Chem Soc 125:2368–2369
116. Zhang Y, Hsung RP, Tracey MR, Kurtz KCM, Vera EL (2004) copper sulfate-pentahydrate-1,10-phenanthroline catalyzed amidations of alkynyl bromides. Synthesis of heteroaromatic amine substituted ynamides. Org Lett 6:1151–1154
117. Zhang X, Zhang Y, Huang J, Hsung RP, Kurtz KCM, Oppenheimer J, Petersen ME, Sagamanova IK, Shen L, Tracey MR (2006) Copper(II)-catalyzed amidations of alkynyl bromides as a general synthesis of ynamides and Z-enamides. An intramolecular amidation for the synthesis of macrocyclic ynamides. J Org Chem 71:4170–4177
118. Dunetz JR, Danheiser RL (2003) Copper-mediated N-alkynylation of carbamates, ureas, and sulfonamides. A general method for the synthesis of ynamides. Org Lett 5:4011–4014
119. Hirano S, Tanaka R, Urabe H, Sato F (2004) Practical preparation of N-(1-Alkynyl)sulfonamides and their remote diastereoselective addition to aldehyde via titanation. Org Lett 6:727–729
120. Laroche C, Li J, Freyer MW, Kerwin SM (2008) Coupling reactions of bromoalkynes with imidazoles mediated by copper salts: synthesis of novel N-alkynylimidazoles. J Org Chem 73:6462–6465
121. Dooleweerdt K, Birkedal H, Ruhland T, Skrydstrup T (2008) Irregularities in the effect of potassium phosphate in ynamide synthesis. J Org Chem 73:9447–9450
122. Burley GA, Davies DL, Griffith GA, Lee M, Singh K (2010) Cu-catalyzed N-alkynylation of imidazoles, benzimidazoles, indazoles, and pyrazoles using PEG as solvent medium. J Org Chem 75:980–983
123. DeKorver KA, Walton MC, North TD, Hsung RP (2011) Introducing a new class of N-phosphoryl ynamides via Cu(I)-catalyzed amidations of alkynyl bromides. Org Lett 13:4862–4865
124. Maligres PE, Krska SW, Dormer PG (2012) A soluble copper(I) source and stable salts of volatile ligands for copper-catalyzed C−X couplings. J Org Chem 77:7646–7651
125. Chen XY, Wang L, Frings M, Bolm C (2014) Copper-catalyzed N-alkynylations of sulfoximines with bromoacetylenes. Org Lett 16:3796–3799
126. Yao B, Liang Z, Niu T, Zhang Y (2009) Iron-catalyzed amidation of alkynyl bromides: a facile route for the preparation of ynamides. J Org Chem 74:4630–4633
127. Nizovtseva TV, Komarova TN, Nakhmanovich AS, Lopyrev VA (2002) Synthesis of 2-acylmethylene-4-amino-1,3-dithien-6-iminium perchilorates. Chem Heterocycl Compd 38:1134–1135
128. Sobenina LN, Demenev AP, Mikhaleva AI, Ushakov IA, Afonin AV, Petrova OV, Elokhina VN, Volkova KA, Tomyashinova DSD, Trofimov BA (2002) Functionally

substituted 1,3-diethyenyl [1,2-c][1, 3]pyrrolothiazoles from pyrrole-2-carbodithioates. Sulfur Lett 25:87–93

129. Glotova TE, Dolgushin GV, Albanov AI, Dvorko MY, Protsuk NI (2005) Reactions of 1-bromo-2-benzoylacetylene with 2,4-dithiobiuretes. J Sulfur Chem 26:359–364

130. Arens JF (1960) In: Raphael RA, Taylor EC, Wynberg H (eds) Advances in organic chemsitry, vol II. Interscience Publishers, Inc., New York, p 117

131. Ziegler GR, Welch CA, Orzech CE, Kikkawa S, Miller SI (1963) Nucleophilic substitution at an acetylenic carbon: acetylenic thioethers from haloalkynes and sodium thiolates. J Am Chem Soc 85:1648–1651

132. Iorga B, Eymery F, Carmichael D, Savignac P (2000) Dialky 1-alkynylphosphonates: a range of promising reagents. Eur J Org Chem 3103–3115

133. Berger O, Petit C, Deal EL, Montchamp JL (2013) Phsphorus-carbon bond formation: palladium-catalyzed cross-coupling of H-phosphinates and other P(O)H-containing compounds. Adv Synth Catal 355:1361–1373

134. Wang Y, Gan J, Liu L, Yuan H, Gao Y, Liu Y, Zhao Y (2014) Cs_2CO_3-promoted one-pot synthesis of alkynylphosphonates,-phosphinates, and -phosphine oxides. J Org Chem 79:3678–3683

135. Wu W, Jiang H (2012) Palladium-catalyzed oxidative of unsaturated hydrocarbons using molecular oxygen. Acc Chem Res 45:1736–1748

136. Huang J, Zhou L, Jiang H (2006) Palladium-catalyzed allylation of alkynes with allyl alcohol in aqueous media: highly regio- and stereoselective synthesis of 1,4-dienes. Angew Chem Int Ed 45:1945–1949

137. Jiang H, Liu X, Zhou L (2008) Firt synthesis of 1-chlorovinyl allenes via palladium-catalyzed allenylation of alkynoates with propargyl alcohols. Chem Eur J 14:11305–11309

138. Gooßen LJ, Rodriguez N, Gooßen K (2009) Stereoselective synthesis of β-chlorovinyl ketones and arenes by the catalytic addition of acid chlorides to alkynes. Angew Chem Int Ed 48:9592–9594

139. Murai M, Hatano R, Kitabata S, Che K (2011) Gallium(III)-catalysed bromocyanation of alkynes: regio- and stereoselective synthesis of β-bromo-α, β-unsaturated nitriles. Chem Commun 47:2375–2377

140. Heasley VL, Buczala DM, Chappell AE, Hill DJ, Whisenand JM, Shellhamer DF (2002) Addition of bromine chloride and iodine monochloride to carbonyl-conjugated, acetylenic ketones: synthesis and mechanisms. J Org Chem 67:2183–2187

141. Bellina F, Colzi F, Mannina L, Rossi R, Viel S (2003) Reaction of alkynes with iodine monochloride revisited. J Org Chem 68:10175–10178

142. Ho ML, Flynn AB, Ogilvie WW (2007) Single-isomer iodochlorination of alkynes and chlorination of alkenes using tetrabutylammonium iodide and dichloroethane. J Org Chem 72:977–983

143. Chen Z, Jiang H, Li Y, Qi C (2010) Highly efficient two-step synthesis of (Z)-2-halo-1-iodoalkenes from terminal alkynes. Chem Commun 46:8049–8051

144. Zhu G, Chen D, Wang Y, Zheng R (2012) Highly stereoselective synthesis of (Z)-1,2-dihaloalkenes by a Pd-catalyzed hydrohalogenation of alkynyl halides. Chem Commun 48:5796–5798

145. Hagan DO (2008) Understanding organofluorine chemistry. An introduction to the C−F bond. Chem Soc Rev 37:308–319

146. Zhang W, Huang W, Hu J (2009) Highly stereoselective synthesis of monofluoroalkenes from α-fluorosulfoximines and nitrones. Angew Chem Int Ed 48:9858–9861

147. Tang P, Wang K, Ritter T (2011) Deoxyfluorination of phenols. J Am Chem Soc 133:11482–11484

148. Chan KSL, Wasa M, Wang X, Yu J (2011) Palladium(II)-catalyzed selective monofluorination of benzoic acids using a practical auxiliary: a weak-coordination approach. Angew Chem Int Ed 50:9081–9084

149. Zhang H, Zhou C, Chen Q, Xiao J, Hong R (2011) Monofluorovinyl tosylate: a useful building block for the synthesis of terminal vinyl monofluorides via suzuki-miyaura coupling. Org Lett 13:560–563
150. Peng H, Liu G (2011) Palladium-catalyzed tandem fluorination and cyclization of enynes. Org Lett 13:772–775
151. Li Y, Liu X, Ma D, Liu B, Jiang H (2012) Silver-assisted difunctionalization of terminal alkynes: highly regio- and stereoselective synthesis of bromofluoroalkenes. Adv Synth Catal 354:2683–2688
152. Akana JA, Bhattacharyya KX, Müller P, Sadighi JP (2007) Reversible C−F bond formation and the Au-catalyzed hydrofluorination of alkynes. J Am Chem Soc 129:7736–7737
153. Amatore C, Jutand A, Duc G (2012) The triple role of fluoride ions in palladium-catalyzed Suzuki-Miyaura reactions: unprecedented transmetalation from [ArPdFL$_2$] complexes. Angew Chem Int Ed 51:1379–1382
154. Donohoe TJ, Butterworth S (2003) A general oxidative cyclization of 1,5-dienes using catalytic osmium tetroxide. Angew Chem Int Ed 42:948–951
155. Zhao J, Chng S, Loh T (2007) Lewis acid-promoted intermolecular acetal-initiated cationic polyene cyclizations. J Am Chem Soc 129:492–493
156. Iwashima M, Terada I, Okamoto K, Iguchi K (2002) Tricycloclavulone and clavubicyclone, novel prostanoid-related marine oxylipins, isolated from the okinawan soft coral *Clavularia viridis*. J Org Chem 67:2977–2981
157. Mondal S, Yadav RN, Ghosh S (2011) Unprecedented influence of remote substituents on reactivity and stereoselectivity in Cu(I)-catalysed [2 + 2] photocycloaddition. An approach towards the synthesis of tricycloclavulone. Org Biomol Chem 9:4903–4913
158. Chen X, Kong W, Cai H, Kong L, Zhu G (2011) Palladium-catalyzed highly regio- and stereoselective synthesis of (1E)-or (1Z)-1,2-dihalo-1,4-dienes via haloallylation of alkynyl halides. Chem Commun 47:2164–2166
159. Chen D, Cao Y, Yuan Z, Cai H, Zheng R, Kong L, Zhu G (2011) Synthesis of cis-1,2-dihaloalkenes featuring palladium-catalyzed coupling of haloalkynes and α, β-unsaturated carbonyls. J Org Chem 76:4071–4074
160. Chen D, Chen X, Lu Z, Cai H, Shen J, Zhu G (2011) Palladium-catalyzed dienylation of haloalkynes using 2,3-butadienyl acetate: a facile access to (1Z)-1,2-dihalo-3-vinyl-1,3-dienes. Adv Synth Catal 353:1474–1478
161. Li J, Yang S, Huang L, Chen H, Jiang H (2013) Highly efficient and practical synthesis of functionalized 1,5-dienes via Pd(II)-catalyzed halohomoallylation of alkynes. RSC Adv 3:11529–11532
162. Kitson RRA, Millemaggi A, Taylor RJK (2009) The renaissance of α-methylene-γ-butyrolactones: new synthetic approaches. Angew Chem Int Ed 48:9426–9451
163. Groll M, Balskus EP, Jacobsen EN (2008) Structural analysis of spiro-β-lactone proteasome inhibitors. J Am Chem Int 130:14981–14983
164. Devalankar DA, Karabal PU, Sudalai A (2013) Optically pure γ-butyrolactones and epoxy esters via two stereocentered HKR of 3-substituted epoxy esters: a formal synthesis of (−)-paroxetine, Ro 67-8867 and (+)-eldanolide. Org Biomol Chem 11:1280–1285
165. Li J, Yang W, Yang S, Huang L, Wu W, Sun Y, Jiang H (2014) Palladium-catalyzed cascade annulation to construct functionalized β- and γ-lactones in ionic liquids. Angew Chem Int Ed 53:7219–7222
166. Li J, Yang S, Jiang H, Wu W, Zhao J (2013) Palladium-catalyzed coupling of alkynes with unactivated alkenes in ionic liquids: a regio- and stereoselective synthesis of functionalized 1,6-dienes and their analogues. J Org Chem 78:12477–12486
167. Li J, Yang S, Wu W, Qi C, Deng Z (2014) Synthesis of 1,4-dienes by Pd(II)-catalyzed haloallylation of alkynes with allylic alcohols in ionic liquids. Tetrahedron 70:1516–1523
168. Zweifel G, Arzoumanian H (1967) α-Halovinylboranes. Their preparation and conversion into *cis*-vinyl halides, *trans*-olefins, ketones, and *trans*-vinylboranes. J Am Chem Soc 89:5086–5088

169. Brown HC, Blue CD, Nelson DJ, Bhat NG (1989) Vinylic organoboranes. 12. Synthesis of (Z)-1-halo-1-alkenes via hydroboration of 1-halo-1-alkynes followed by protonolysis. 54:6064–6067

170. Salvi L, Jeon S, Fisher EL, Carroll PJ, Walsh PJ (2007) Catalytic asymmetric generation of (Z)-disubstituted allylic alcohols. J Am Chem Soc 129:16119–16125

171. Tsuji H, Fujimoto T, Endo K, Nakamura M, Nakamura E (2008) Stereoselective synthesis of trisubstituted E-iodoalkenes by indium-catalyzed syn-addition of 1,3-dicarbonyl compounds to 1-iodoalkynes. Org Lett 10:1219–1221

172. Li Y, Zhao J, Chen H, Liu B, Jiang H (2012) Pd-catalyzed and CsF-promoted reaction of bromoalkynes with isocyanides: regioselective synthesis of substituted 5-iminopyrrolones. Chem Commun 48:3545–3547

173. Aparece MD, Vadola PA (2014) Gold-catalyzed dearomative spirocyclization of aryl alkynoate esters. Org Lett 16:6008–6011

174. Lucette D, Marie PJ (1979) β-Lithhioenamines from β-chloroenmanie. A convenient preparation method for hindered ketones. J Org Chem 44:3585–3586

175. Campos PJ, Tan C, Rodríguez MA, Añón E (1996) Preparation of 3-haloquinolines from 3-amino-2-halo-2-alkenimines. J Org Chem 61:7195–7197

176. Ji X, Huang H, Wu W, Jiang H (2013) Palladium-catalyzed intermolecular dehydrogenative aminohalogenation of alkenes under molecular oxygen: an approach to brominated enamines. J Am Chem Soc 135:5286–5289

177. Zhang J, Meng L, Li P, Wang L (2013) The sequential reactions of tetrazoles with bromoalkynes for the synthesis of (Z)-N-(2-bromo-1-vinyl)-N-arylcyanamides and 2-arylindoles. RSC Adv 3:6807–6812

178. Kowalski CJ, O'Dowd ML, Burker MC, Fields KW (1980) alpha.-Keto dianions. New reactive intermediates. J Am Chem Soc 102:5411–5412

179. Kowalski CJ, Haque MS, Fields KW (1985) Ester homologation via.alpha.-bromo.alpha.-keto dianion rearrangement. J Am Chem Soc 107:1429–1430

180. Kowalski CJ, Haque MS (1985) Bromomethyl ketones and enolates: alternative products from ester homologation reactions. J Org Chem 50:5140–5148

181. Pincock JA, Yates K (1970) Kinetics and mechanism of electrophilic bromination of acetylenes. Can J Chem 48:3332–3348

182. Barluenga J, Rodriguez MA, Campos PJ (1990) Electrophilic additions of positive iodine to alkynes through an iodonium mechanism. J Org Chem 55:3104–3106

183. Chen Z, Li J, Jiang H, Zhu S, Li Y, Qi C (2010) Silver-catalyzed difunctionalization of terminal alkynes: highly regio- and stereoselective synthesis of (Z)-γ-haloenol acetates. Org Lett 12:3262–3265

184. Chary BC, Kim S, Shin D, Lee PH (2011) A regio- and stereoselective synthesis of trisubstituted alkenes via gold(I)-catalyzed hydrophosphoryloxylation of haloalkynes. Chem Commun 47:7851–7853

185. Chen X, Chen D, Lu Z, Kong L, Zhu G (2011) Palladium-catalyzed coupling of haloalkynes with allyl acetate: a regio- and stereoselective synthesis of (Z)-β-haloenol acetates. J Org Chem 76:6338–6343

186. Xie L, Wu Y, Yi W, Zhu L, Xiang J, He W (2013) Gold-catalyzed hydration of haloalkynes to α-halomethyl ketones. J Org Chem 78:9190–9195

187. Chen Z, Ye D, Ye M, Zhou Z, Li S, Liu L (2014) AgF/TFA-promoted highly efficient synthesis of α-haloketones from haloalkynes. Tetrahedron Lett 55:1373–1375

188. Xu H, Gu S, Chen Z, Li D, Dou J (2011) TBAF-mediated reactions of 1,1-dibromo-1-alkenes with thiols and amines and regioselective synthesis of 1,2-heterodisubstituted alkenes. J Org Chem 76:2448–2458

189. Liu G, Kong L, Shen J, Zhu G (2014) A regio- and stereoselective entry to (Z)-β-halo alkenyl sulfides and their applications to access stereodefined trisubstituted alkenes. Org Biomol Chem 12:2310–2321

190. Wender PA, Gamber GG, Hubbard RD, Zhang L (2002) Three-component cycloadditions: the first transiton metal-catalyzed [5 + 2 + 1] cycloaddition reactions. J Am Chem Soc 124:2876–2877
191. Feng J, Zhang J (2011) An atom-economic synthesis of bicyclo[3.1.0]hexanes by Rhodium N-heterocyclic carbene-catalyzed diastereoselective tandem hetero-[5 + 2] cycloaddition/claisen rearrangement reaction of vinylic oxiranes with alkynes. J Am Chem Soc 133:7304–7307
192. Misudo T, Naruse H, Kondo T, Ozaki Y, Wantanabe Y (1994) [2 + 2]Cycloaddition of norbornenes with alkynes catalyzed by ruthenium complexes. Angew Chem Int Ed 33:580–581
193. Villeneuve K, Tam W (2004) Asymmetric induction in ruthenium-catalyzed [2 + 2] cycloadditions between bicyclic alkenes and a chiral acetylenic acyl sultam. Angew Chem Int Ed 43:610–613
194. Villeneuve K, Riddell N, Jordan RW, Tsui GC, Tam W (2004) Ruthenium-catalyzed [2 + 2] cycloadditions between bicyclic alkenes and alkynyl halides. Org Lett 6:4543–4546
195. Jordan RW, Villeneuve K, Tam W (2006) Study on the reactivity of the alkyne component in ruthenium-catalyzed [2 + 2] cycloadditions between an alkene and an alkyne. J Org Chem 71:5830–5833
196. Allen A, Villeneuve K, Cockburn N, Fatila E, Riddell N, Tam W (2008) Alkynyl halides in ruthenium(II)-catalyzed [2 + 2] cycloadditions of bicyclic alkenes. Eur J Org Chem 2008:4178–4192
197. Tsvetkov NP, Koldobskii AB, Godovikov IA, Kalinin VN (2005) [2 + 2] Cycloaddition reactions of 1-trifluoroacetyl-2-chloroacetylene with vinyl ethers. Dokl Chem 404:210–211
198. Koldobskii AB, Tsvetkov NP, Godovikov IA, Kalinin VN (2009) [2 + 2] Cycloaddition and electrophilic alkynylation reactions of 4-chloro-1,1,1-trifluorobut-3-yn-2-one and alkyl vinyl ethers. Russ Chem Bull Int Ed 58:1438–1440
199. Aikawa K, Hioki Y, Shimizu N, Mikami K (2011) Catalytic asymmetric synthesis of stable oxetenes via lewis acid-promoted [2 + 2] cycloaddition. J Am Chem Soc 133:20092–20095
200. Kolb HC, Finn MG, Sharpless KB (2001) Click chemistry: diverse chemical function from a few good reactions. Angew Chem Int Ed 40:2004–2021
201. Hein JE, Tripp JC, Krasnova LB, Sharpless KB, Fokin VV (2009) Copper(I)-catalyzed cycloaddition of organic azides and 1-iodoalkynes. Angew Chem Int Ed 48:8018–8021
202. García-Álvarez J, Díez J, Gimeno J (2010) A highly efficient copper(I) catalyst for the 1,3-dipolar cycloaddition of azides with terminal and 1-iodoalkynes in water: regioselective synthesis of 1,4-disubstituted and 1,4,5-trisubstituted 1,2,3-triazoles. Green Chem 12:2127–2130
203. Juríček M, Stout K, Kouwer PHJ, Rowan AE (2011) Fusing triazoles: toward extending aromaticity. Org Lett 13:3494–3497
204. Brotherton WS, Clark RJ, Zhu L (2012) Synthesis of 5-Iodo-1,4-disubstituted-1,2,3-triazoles mediated by in situ generated copper(I) catalyst and electrophilic triiodide ion. J Org Chem 77:6443–6455
205. Lal S, Rzepa HS, Díez-González (2014) Catalytic and computational studies of N-heterocyclic carbene or phosphine-containing copper(I) complexes for the synthesis of 5-iodo-1,2,3-triazoles. ACS Catal 4:2274–2287
206. Rostovtsev VV, Green LG, Fokin VV, Sharpless KB (2002) A stepwise huisgen cycloaddition process: copper(I)-catalyzed regioselective "Ligation" of azides and terminal alkynes. Angew Chem Int Ed 41:2596–2599
207. Himo F, Lovell T, Hilgraf R, Rostovtsev VV, Noodleman L, Sharpless KB, Fokin VV (2005) Copper(I)-catalyzed synthesis of azoles. DFT study predicts unprecedented reactivity and intermediates. J Am Chem Soc 127:210–216
208. Talley JJ, Bertenshaw ST, Brown DL, Carter JS, Graneto MJ, Kellogg MS, Koboldt CM, Yuan J, Zhang Y, Seibert K (2000) N-[[(5-Methyl-3-phenylisoxazol-4-yl)- phenyl]sulfonyl] propanamide, sodium salts, parecoxib sodium: a potent and selective inhibitor of COX-2 for parenteral administration. J Med Chem 43:1661–1663

209. Habbe AG, Rao PNP, Knaus EE (2001) Design and synthesis of 4,5-diphenyl-4-isoxazolines: novel inhibitors of cylcooxygenase-2 with analgesic and antiinflammatory activity. J Med Chem 44:2921–2927

210. Kotali E, Varvoglis A, Bozopoulo A (1989) Reactivity of arylethynyl(phenyl)iodonium salts towards some 1,3-dipoles. X-Ray molecular structure of 3-mesityl-5-phenylisoxazol-4-yl (phenyl)iodonium toluene-p-sulphonate. J Chem Soc Perkin Trans 1:827–832

211. Letourneau JJ, Riviello C, Ohlmeyer MHJ (2007) A novel and convenient synthesis of 5-aryl-4-bromo-3-carboxyisoxazoles: useful intermediates for the solid-phase synthesis of 4,5-diarylisoxazoles. Tetrahedron Lett 48:1739–1943

212. Crossley JA, Browne DL (2010) An alkynyliodide cycloaddition strategy for the construction of iodoisoxazoles. J Org Chem 75:5414–5416

213. Hanson SM, Morlock EV, Satyshur KA, Czajkowski C (2008) Structural requirements for eszopiclone and zolpidem binding to the γ-aminobutyric acid type-A (GABA$_A$) receptor are different. J Med Chem 51:7243–7252

214. Singhaus RR, Bernotas RC, Steffan R, Matelan E, Quinet E, Nambi R, Feingold I, Huselton C, Wilhelmsson A, Goos-Nilsson A, Wrobel J (2010) 3-(3-Aryloxyaryl)imidazo [1,2-a]pyridine sulfones as liver X receptor agnists. Bioorg Med Chem Lett 20:521–525

215. Gao Y, Yin M, Wu W, Huang L, Jiang H (2013) Copper-catalyzed intermolecular oxidative cyclization of haloalkynes: synthesis of 2-halo-substituted imidazo[1,2-a]pyridines, Imidazo [1,2-a]pyrazines and Imidazo[1,2-a]pyrimidines. Adv Synth Catal 355:2263–2273

216. Trofimov BA, Sobenina LN, Demenv AP, Mikhaleva AI (2004) C-vinylpyrroles as pyrrole building blocks. Chem Rev 104:2481–2506

217. Fan H, Peng J, Hamann MT, Hu J (2008) Lamellarins and related pyrrole-derived alkaloids from marine organisms. Chem Rev 108:264–287

218. Gulevich AV, Dudnik AS, Chernyak N, Gevorgyan V (2013) Transition metal-mediated synthesis of monocyclic aromatic heterocycles. Chem Rev 113:3084–3213

219. Sai M, Matsubara S (2011) Silver-catalyzed intramolecular chloroamination of allenes: easy access to functionalized 3-pyrroline and pyrrole derivatives. Org Lett 13:4676–4679

220. Debrouwer W, Heugebaert TSA, Stevens CV (2014) Preparation of tetrasubstituted 3-phosphonopyrroles through hydroamination: scope and limitations. J Org Chem 79:4322–4331

221. Zheng J, Huang L, Li Z, Wu W, Li J, Jiang H (2015) Synthesis of 3-bromosubstituted pyrroles via palladium-catalyzed intermolecular oxidative cyclization of bromoalkynes with N-allyamines. Chem Commun 51:5894–5897

222. Wang L, Huang J, Peng S, Liu H, Jiang X, Wang J (2013) Palladium-catalyzed oxidative cycloaddition through C—H/N—H activation: access to benzazepines. Angew Chem Int Ed 52:1768–1772

223. Li W, Liu C, Zhang H, Ye K, Zhang G, Zhang W, Duan Z, You S, Lei A (2014) Palladium-catalyzed oxidative carbonylation of N-allylamines for the synthesis of β-lactams. Angew Chem Int Ed 53:2443–2446

224. Wender PA, Jenkins TE, Suzuki S (1995) Transition metal-catalyzed intramolecular [4 + 2] diene-allene cycloadditions: a convenient synthesis of angularly substituted ring systems with provision for catalyst-controlled chemo- and stereocomplementarity. J Am Chem Soc 117:1843–1844

225. Francos J, Grande-Carmona F, Faustino H, Lglsesias-Sigüenza J, Díez E, Alonso I, Fernández R, Lassaletta JM, López F, Mascareñas JL (2012) Axially chiral triazoloisoquinolin-3-ylidene ligands in Gold(I)-catalyzed asymmetric intermolecular (4 + 2) cycloadditions of allenamides and dienes. J Am Chem Soc 134:14322–14325

226. Zhang C, Trudell ML (1996) A short and efficient total synthesis of (±)-epibatidine. J Org Chem 61:7189–7191

227. Yoo W, Allen A, Villeneuve K, Tam W (2005) rhodium-catalyzed intramolecular [4 + 2] cycloadditions of alkynyl halides. Org Lett 7:5853–5856

228. Iwayama T, Sato Y (2009) Nickel(0)-catalyzed[2 + 2 + 2]cycloaddition of diynes and 3,4-pyridynes: novel synthesis of isoquinoline derivative. Chem Commun 35:5245–5247

229. Nicolaou KC, Tang Y, Wang J (2009) Total synthesis of sporolide B. Angew Chem Int Ed 48:3449–3453
230. Katritzky AR (2004) Introduction: heterocycles. Chem Rev 104:2125–2126
231. Fukudome Y, Naito H, Hata T, Urabe H (2008) Copper-catalyzed 1,2-double amination of 1-halo-1-alkynes. Consise synthesis of protected tetrahydropyrazines and related heterocyclic compounds. J Am Chem Soc 130:1820–1821
232. Naito H, Hata T, Urabe H (2008) Facile preparation of N-protected 2-alkylidene-1,3-imidazolidines. Tetrahedron Lett 49:2298–2301
233. Chen Z, Zeng W, Jiang H, Liu L (2012) Cu(II)-catalyzed synthesis of naphthalene-1,3-diamine derivatives from haloalkynes and amines. Org Lett 14:5385–5387
234. Chen Z, Jiang H, Pan X, He Z (2011) Practical synthesis of amides from alkynyl bromides, amines, and water. Tetrahedron 67:5920–5927
235. Sun Y, Jiang H, Wu W, Zeng W, Li J (2014) Synthesis of thioamides via one-pot A^3-coupling of alkynyl bromides, amines, and sodium sulfide. Org Biomol Chem 12:700–707
236. Horton DA, Bourne GT, Smythe ML (2003) The combinatorial synthesis of bicyclic privileged structures or privileged substructures. Chem Rev 103:890–893
237. Alonso F, Beletskaya IP, Yus M (2004) Transition-metal-catalyzed addition of heteroatom-hydrogen bonds to alkynes. Chem Rev 104:3079–3160
238. Zeni G, Larock RC (2006) Synthesis of heterocycles via palladium-catalyzed oxidative addition. Chem Rev 106:4644–4680
239. Wang S, Li P, Yu L, Wang L (2011) Sequential and one-pot reactions of phenols with bromoalkynes for the synthesis of (Z)-2-bromovinyl phenyl ethers and benzo[b]furans. Org Lett 13:5968–5971
240. Zeng W, Wu W, Jiang H, Sun Y, Chen Z (2013) Highly efficient synthesis of 2,3,4-trisubstituted furans via silver-catalyzed sequential nucleophilic addition and cyclization reactions of haloalkynes. Tetrahedron Lett 54:4605–4609
241. Yoshida H, Shirakawa E, Honda Y, Hiyama T (2002) Addition of ureas to arynes: straightforward synthesis of benzodiazepine and benzodiazocine derivatives. Angew Chem Int Ed 114:3247−3249
242. Dutton FE, Lee BH, Johnson SS, Coscarelli EM, Lee PH (2003) Restricted conformation analogues of an anthelmintic cyclodepsipeptide. J Med Chem 46:2057–2073
243. Geronikaki AA, Dearden JC, Filimonov D, Galaeva I, Garibova TL, Gloriozova T, Kraneva V, Lagunin A, Macaev FZ, Molodavkin G, Poroikov VV, Pogrebnoi SI, Shepeli F, Voronina TA, Tsitlakidou M, Vlad L (2004) Design of new cognition enhancers: from computer prediction to synthesis and biological evaluation. J Med Chem 47:2870–2876
244. Liu B, Li Y, Yin M, Wu W, Jiang H (2012) Palladium-catalyzed tandem reaction of o-aminophenols, bromoalkynes and isocyanides to give 4-amine-benzo[b][1,4]oxazepines. Chem Commun 48:11446–11448
245. Peng J, Shang G, Chen C, Miao Z, Li B (2013) Nucleophilic addition of benzimidazoles to alkynyl bromides/palladium-catalyzed intramolecular C−H vinylation: synthesis of benzo [4,5]imidazo[2,1-a]isoquinolines. J Org Chem 78:1242–1248
246. Jiang H, Yin M, Li Y, Liu B, Zhao J, Wu W (2014) An efficient synthesis of 2,5-diimino-furans via Pd-catalyzed cyclization of bromoacrylamides and isocyanides. Chem Commun 50:2037–2039

Chapter 4
Conclusions and Outlook

Abstract The above three chapters have highlighted the robust reactivity and described the general preparation methods of haloalkyne reagents. Significantly, efforts have been made in elucidating the mechanisms of these chemical processes, which provide the researchers valuable insight of the haloalkyne compounds. In this chapter, we will summarize the book and also point out some challenges that need to be faced in the area of haloalkyne chemistry. We hope this book will not only ignite the interest of readers to the field of haloalkyne chemistry, but also inspire the researchers to answer the unsolved challenges and exploit new research areas of haloalkyne chemistry.

Keywords Insight and understanding · Unsolved challenges · Haloalkyne chemistry

This book has highlighted the robust reactivity and also summarized the general preparation methods of haloalkyne reagents. The diverse reactivity of haloalkynes allow these efficient transformations to deliver a variety of novel acyclic and cyclic structures representing prevalent and significant frameworks as well as being useful motifs for further transformations. Thus, haloalkynes have emerged as powerful and versatile building blocks in a diverse spectrum of synthetic transformations including natural product total synthesis. In addition, the efforts have been made in elucidating the mechanisms of these chemical processes, which provide the researchers valuable insight and understanding the reactivity of haloalkyne compounds.

Despite the great progress that has achieved, many challenges still need to be faced in the area of haloalkyne chemistry, such as, the catalytic asymmetric reactions and the carbon nucleophilic addition reactions of haloalkynes, as well as the development of more general, practical, efficient and green methods for the transformation of haloalkynes to access valuable molecules. We hope this book will not only ignite the interest of readers to the field of haloalkyne chemistry, but also inspire the researchers to answer the unsolved challenges and exploit new research areas of haloalkyne chemistry.

© The Author(s) 2016
H. Jiang et al., *Haloalkyne Chemistry*, SpringerBriefs in Green Chemistry for Sustainability, DOI 10.1007/978-3-662-49001-3_4

Printed in the United States
By Bookmasters